Stochastic Analysis of Computer Storage

Mathematics and Its Applications

Stochastic Analysis of Computer Storage

by

O. I. Aven
Moscow Institute of the Food Industry, Moscow, U.S.S.R.

E. G. Coffman, Jr.
AT&T Bell Laboratories, Murray Hill, New Jersey, U.S.A.

and

Y. A. Kogan
Institute of Control Sciences, Moscow, U.S.S.R.

D. Reidel Publishing Company

A MEMBER OF THE KLUWER ACADEMIC PUBLISHERS GROUP

Dordrecht / Boston / Lancaster / Tokyo

Library of Congress Cataloging in Publication Data

Aven, Oleg Ivanovich.
  Stochastic analysis of computer storage.

  (Mathematics and its applications)
  Includes index.
  1. Computer storage devices—Mathematical models. 2.   Stochastic analysis.
  I. Coffman, E. G. (Edward Grady), 1934–   II.   Kogan, Y. A. (Yakov
Afroimovich).   III.   Title.   IV.   Series: Mathematics and its applications
(D. Reidel Publishing Company)
TK7895.M4A94   1987   621.397'0724   87–12821
ISBN 90–277–2515–2

Published by D. Reidel Publishing Company,
P.O. Box 17, 3300 AA Dordrecht, Holland.

Sold and distributed in the U.S.A. and Canada
by Kluwer Academic Publishers,
101 Philip Drive, Assinippi Park, Norwell, MA 02061, U.S.A.

In all other countries, sold and distributed
by Kluwer Academic Publishers Group,
P.O. Box 322, 3300 AH Dordrecht, Holland.

Distribution rights for the COMECON countries by
Scientific Information Consultants,
661 Finchley Road
London NW 2 2HN

Printed in The Netherlands

SERIES EDITOR'S PREFACE

Growing specialization and diversification have brought a host of monographs and textbooks on increasingly specialized topics. However, the "tree" of knowledge of mathematics and related fields does not grow only by putting forth new branches. It also happens, quite often in fact, that branches which were thought to be completely disparate are suddenly seen to be related.

Further, the kind and level of sophistication of mathematics applied in various sciences has changed drastically in recent years: measure theory is used (non-trivially) in regional and theoretical economics; algebraic geometry interacts with physics; the Minkowsky lemma, coding theory and the structure of water meet one another in packing and covering theory; quantum fields, crystal defects and mathematical programming profit from homotopy theory; Lie algebras are relevant to filtering; and prediction and electrical engineering can use Stein spaces. And in addition to this there are such new emerging subdisciplines as "experimental mathematics", "CFD", "completely integrable systems", "chaos, synergetics and large-scale order", which are almost impossible to fit into the existing classification schemes. They draw upon widely different sections of mathematics. This programme, Mathematics and Its Applications, is devoted to new emerging (sub)disciplines and to such (new) interrelations as exempla gratia:

- a central concept which plays an important role in several different mathematical and/or scientific specialized areas;
- new applications of the results and ideas from one area of scientific endeavour into another;
- influences which the results, problems and concepts of one field of enquiry have and have had on the development of another.

The Mathematics and Its Applications programme tries to make available a careful selection of books which fit the philosophy outlined above. With such books, which are stimulating rather than definitive, intriguing rather than encyclopaedic, we hope to contribute something towards better communication among the practitioners in diversified fields.

New technology generates new mathematical problems. After all, as the David report remarked, the currently so-celebrated high technology is essentially mathematical technology. Many of these problems can, of course, be handled by employing or adapting more-or-less standard techniques. Occasionally however - and this seems to happen more frequently of late - the problems ask completely different questions than ever before and call for new and unexplored mathematical models. Two recent examples are provided by flexible manufacturing systems and (efficient) allocation problems in computers. Actually there are two such problems (at least): allocation of processors, a set of problems falling well into the established field of queueing theory and allocation of memory space. The latter presents many new aspects, e.g. involving spatial phenomena such as fragmentation, which is notorious for slowing down things. As every computer buff knows, memory, or more precisely, efficient use of memory storage space, remains a vastly important matter. This will also remain so.

Thus a new young and vigorous area of research has sprung up together with the usual host of heuristic algorithms and principles demanding systematic study and evaluation, the development of the mathematics to do so, and research monographs presenting matters in a coherent and integrated fashion. That is precisely what this book does for (stochastic) models for computer memory storage allocation.

The unreasonable effectiveness of mathematics in science ...

    Eugene Wigner

Well, if you know of a better 'ole, go to it.

    Bruce Bairnsfather

What is now proved was once only imagined.

    William Blake

As long as algebra and geometry proceeded along separate paths, their advance was slow and their applications limited.

But when these sciences joined company they drew from each other fresh vitality and thenceforward marched on at a rapid pace towards perfection.

    Joseph Louis Lagrange.

Bussum, April 1987   Michiel Hazewinkel

Table of Contents

Chapter 1

Background and Overview

1. Introduction

Storage and processing time are the two major resources of computer systems; it is in these terms that the performance of computers is most often evaluated. In many applications the individual requirements for these resources are interdependent. For example, by re-designing algorithms it is frequently possible to increase computing speeds by making more storage available. By the same token, depending on the application, either resource may appear as the chief limitation, or bottleneck, to the "size" of the problem that can be solved. Since these limitations promise to continue in spite of technological advances, we have an obvious motivation for the study of efficient processor and storage allocation techniques and the analysis of their performance.

Over the past 20 years the extensive research in methods of processor allocation has brought this aspect of computer operation to a point where it is relatively well understood. However, the mathematical models analyzed have been simplistic, because the critical influence of storage requirements has largely been ignored. In contrast, while there have been a number of significant contributions, the analysis of models of storage allocation, and more importantly models combining both storage and processor allocation, is still a young and developing research area.

In this work we focus on the analysis of processes arising in models of storage allocation. More precisely, we concentrate on stochastic models in which there are sequences of "arrivals" and "departures" in a realistic dynamic setting. The arrivals may correspond directly to requests for storage which subsequently depart in the sense that the space allocated to them is again made available to other requests. The correspondence to requests may also be implicit as in sequences of references to information, e.g. program data, where a request for storage space is associated with any reference to information not currently allocated primary storage.

Limited textual material on the subject of computer storage problems has appeared in the past (see [CofD, GelM], for example) in books dealing with the performance analysis of computer systems. Our exclusive treatment is much more extensive in its presentation of up-to-date results and in its bringing together much of the eastern and western literatures on the analysis of storage allocation.

In spite of the importance of the problems, the stochastic analysis of storage allocation has been relatively slow to develop. The obvious reason is that the problems posed have been very difficult to solve. It has been an area of research where the past in other fields of applied mathematics has provided few ready-made results. To models of processor allocation, which have been well served by the classical theory of queues, another dimension has had to be added; units in storage models (e.g. records, files, pages) are described by both a residence (processing) time and a storage requirement.

A closer look reveals several detailed effects that render storage models more difficult. For example, the service mechanism is often compounded of two processes: (1) the access mechanism whereby the relative, physical positioning of information and read/write devices is accomplished, and (2) the actual read/write or information transfer process itself. As another example, the sizes of items are usually preserved while they are in the system, i.e. the sizes of departing items can not be resamplings from a given size distribution, as in the theory of dams for example [Gan, GanP, Ken]. Other effects that we shall be analyzing are finite capacities and the fragmentation (the pattern of alternating occupied and unoccupied areas) that develops as items of varying sizes come and go in storage. The general result of these effects is to make it very difficult to formulate tractable Markov processes describing storage occupancy; the corresponding state spaces and transition functions become too large and unwieldy.

Storage can be classified according to both physical devices and the purposes they serve. Our interest centers on the latter classification. In chapters 2 and 3 we study basic storage processes in buffer and primary memories; in chapters 4 and 5 we investigate the problems of managing the interaction between primary and secondary memories with an emphasis on paging systems; and in chapter 6 we analyze generic models of secondary (auxiliary) storage devices. Because of the generality of our models we shall have no need for detailed characteristics of storage devices. However, the reader will be assumed familiar with the basic dynamics of operation of the more important storage devices. Introductory descriptions of such devices can be found in any introductory text on computer architecture.

The research on storage systems not covered by this book is, in the main, concerned with either archival storage problems or the analysis of combinatorial models. In archival storage records retain their positions indefinitely. Typical archival problems correspond to (1) techniques for optimally assigning records to locations within a storage device based on given access frequencies so as to minimize expected access times; and (2) algorithms for packing a set of records having a given size distribution into multiple storage units (e.g. disk cylinders) so as to minimize the expected number of units used. An extensive treatment of the problems of the first type can be found in a recent text by Wong [Won]. In [CofGJ] the reader will find a broad survey of the latter research under the name of bin-packing. While the models of these interesting problems often include probabilistic assumptions, the results for the most part entail combinatorial rather than stochastic analysis.

Queueing theory and its applications provide important background for stochastic models of storage. Indeed, when items are all assumed to have the same size an analysis of the *number* of items stored, which is a classical process of interest in queueing theory, often becomes mathematically equivalent to an analysis of the *space* occupied by the items. For these cases the reader will find the appropriate analysis in the queueing literature.

On the other hand, we shall have a number of occasions to assume equal-size items when the resulting model does not correspond to a classical queueing model. Such cases will arise because of a number of features peculiar to the storage systems discussed, e.g. loading/unloading operations (chapter 2), fragmentation (section 3.6), replacement control in finite memories (chapter 4 and 5) and non-standard sequencing rules (chapter 6). In fact, the engineering decision to adopt equal item sizes is usually made in the interest of dealing more effectively with these aspects of storage systems. Prominent examples are fixed size *packets* in computer/communication networks, and the fixed size *pages* in computer systems.

In the remainder of this chapter we briefly preview the remaining chapters. Although subsequent chapters are self-sufficient, the reader should find the discussion below helpful in acquiring a coherent view of the book as a whole.

Except for chapters 4 and 5 the chapters are largely independent with little need for cross referencing. This independence carries over to notation as well. The reader should be alerted to the fact that, because of the very large, collective need for notation, the selections made in one chapter are not always consistent with those made in the others. That is, a given symbol may have different meanings in different chapters. Also, use of two dimensions in referring to equations and subsections has been minimized. For example, (2.3) refers to equation (3) of chapter 2, but such a reference can appear only in a chapter other than chapter 2; in chapter 2 the reference will simply be (3). A similar rule applies to section referencing.

A final word on terminology is appropriate. In an attempt to improve readability we have made free use of synonymous words. Important examples are the synonyms *memory* and *storage* (or simply *store*) and the synonyms *records*, *files* or more generally, *items*. Other examples will be noted in context, where confusion might otherwise occur.

2. Fundamental Storage Models

In chapters 2 and 3 we examine a number of basic models of storage systems, each treated largely as an isolated device. In chapter 2 the focus is on buffer systems in which a storage device forms the communication interface between input and output (source and sink) processes. While in general the input and output processes may be independently driven (see for example the producer/consumer systems described in [CofD, Hans]), the primary applications of the models in chapter 2 are to computer/communication settings in which information is received at one rate from a given source and retransmitted at another rate to a given sink. The speed mismatch is commonly brought about by the hardware differences in two communication sub-systems.

In the first model of chapter 2 a single source-process consists of alternating transmission and idle periods. The sizes of transmitted items are i.i.d. (independently and identically distributed) with a general distribution function. The idle periods are assumed to be independently and exponentially distributed with expected value $1/\lambda$. The information received during a source transmission period defines an *item* (message, file, etc.) whose size is proportional to the time required to accumulate it in the buffer. The constant of proportionality is the source transmission rate, r. Items are subsequently retransmitted at a constant rate which defines the output channel. The principal objective of the analysis is the buffer-occupancy process, i.e. the amount of the buffer occupied by items as a function of time. A variety of different operational modes are analyzed; these modes are defined by answers to such questions as:

1. When an item arrives at an empty buffer can its transmission on the output channel begin immediately or must it wait for the loading of the entire item into the buffer?

2. Do items begin loading into a buffer immediately or are they accumulated bit by bit into a register whose contents are transferred in parallel (bit wise) to the buffer?

3. What protocols are in effect when overflows occur in a finite capacity buffer? For example, are items which would cause overflow rejected in their entirety, or just those portions of them that do not fit?

As with the other models of chapters 2 and 3 the important design question addressed by the analysis is the buffer size needed to assure an acceptably low frequency of overflow events.

The material of chapter 2 on single-source buffer systems is based on the recent work of Beneš [Ben1]. Earlier, related research, which we shall not cover in detail, can also be found on "instantaneous loading" models, i.e. the limiting case $r \rightarrow \infty$, where the input becomes a Poisson process of items with a general distribution function governing their sizes. For example, if the buffer is assumed to be infinite, an analysis leads to the classical Takàcs equation [Tak]. We shall see that certain forms of this result arise in the limit $r \rightarrow \infty$ of expressions in section 2.

Gaver and Miller [GavM] and Benês [Ben1] have generalized the Takàcs equation to buffers with a finite capacity. Related questions concerning this type of model are studied in [ArrKS, Gan, GanP, Ken].

Gaver and Miller have also extended the analysis by considering buffers with output rates $r_0(x)$ that are functions of the buffer level x. Their chief example of such a system specifies distinct output rates depending on whether the current buffer content falls above or below a given value of R. Such an assumption does not change the Markov property of the buffer-content process, but it leads to considerably more complex expressions for the transforms of stationary distributions. As in the other research cited the analytical methods are very similar to those described in chapter 2. Note that this model has potential application in those buffer systems consisting of two levels of storage, the faster level having capacity R.

The second model analyzed in chapter 2 extends to systems with a finite number $N > 1$ of sources alternating independently between idle and transmission periods as in the first model. The cost of this added generality lies in the limitations to an infinite buffer size and exponentially distributed transmission periods as well as idle periods. Our analysis follows closely that of Anick, Mitra and Sondhi [AniMS], who extended the earlier work of Kosten [Kos2] on the limiting case with $N \rightarrow \infty$, $\lambda \rightarrow 0$ so that the product of the arrival rate and the average effective service time (viz. the traffic intensity) remains finite. The model and certain other variants of it have been studied elsewhere as well [ChuL,Coh2,GavL].

The models of chapter 3 represent storage processes in primary, random access computer memories. In this setting items have random sizes and correspond to programs and data. Requests to store items arrive over time; once stored, items remain in memory for time periods (residence times) governed by a given probability distribution. At the end of its residence time an item "departs" leaving the space it occupied available for other items.

In section 3.2 the storage system is modeled as an infinite capacity queue with Poisson arrivals, a general service time distribution and service in the order of arrival. The sizes and service times of items are described by a general, joint distribution function and the objective of the analysis is the stationary distribution of the total space occupied by items in the system. The analysis of this section originated in a paper by Sengupta [Sen].

This space occupancy distribution is also the objective of the next three sections, which deal with storage systems having finite capacities. Our results here are taken from a recent paper by Beneš [Ben1]. Implicitly, total occupied space as a performance measure ignores the effects of storage fragmentation, i.e. the alternating occupied and unoccupied regions that build up as items come and go. We return shortly to a discussion of systems for which fragmentation is an important issue.

The analysis in sections 3.2-3.6 illustrates clearly the inherent difficulties of storage models that must obey a principle of conservation of item sizes; i.e. the sequence of item sizes in a sample function of the departure process must be a permutation of the item sizes in the corresponding sample function of the arrival process. This contrasts, for example, with the theory of dams, where it is commonly assumed that the amounts at the output of a reservoir depend on the level of the reservoir, but not on the individual amounts by which this level was established at the input.

The principle of conservation in sections 3.2-3.6 has the immediate result that, even though our chief interest is only in the total amount of memory occupied, the states in any Markov model of a finite memory with arbitrary item sizes must include the size of each item in memory. As is to be expected the added dimensionality complicates the analysis and leads to awkward results, except in special cases.

Chapter 3 next turns to stochastic models of storage fragmentation. The general model studied includes the arrival and departure mechanisms already defined and the storage capacity is assumed to be infinite. Here, however, the method of allocating space to arriving items is made explicit and is subject to the following assumptions.

1.   Items are allocated contiguous space at the time of arrival, i.e. there is no pro-

vision for queueing external to memory, and items cannot be broken up and distributed in any way.

2.   An item once allocated cannot be moved prior to its departure.

It follows from these assumptions that holes of unused space develop as items come and go over time. Thus, although the space required by items currently in memory is simply the sum of their sizes, the total space devoted to (spanned by) these items must also include the cumulative size of the interior holes.

The formation and subsequent utilization of holes in storage is determined by the allocation algorithm in use. Here, we study the basic first-fit algorithm: each arriving item is placed into the first hole found in a linear scan of memory which is large enough to accommodate the item. An exact analysis is provided for a Markov model of the special case when all item sizes are equal, and simulation results are quoted for more general assumptions. In contrast to earlier sections the objective is the distribution of the total space spanned by items currently in memory, including the cumulative size of the interior holes. The analysis follows closely the work of Coffman, Kadota and Shepp [CofKS1]. An interesting result suggested by their analysis and simulation data is that first-fit is asymptotically optimal in the following sense: the fraction of the expected total space spanned which is "wasted" in the interior holes tends to zero as the traffic intensity (which in this model is also the average number of items present) tends to infinity.

Chapter 3 concludes with the study of continuous-path models of the storage process analyzed in sections 3.2 and 3.3, i.e. when storage fragmentation is not an issue. Within these models diffusion approximations of the storage occupancy process are introduced, based on the work of Coffman and Reiman [CofR]. This storage process is analyzed under two different regimes, each assuming that arrivals constitute a general renewal process: a single-server system with general service times, and an infinite server system with Markov (exponential) service times. In the first case, reflected Brownian motion is exhibited as a heavy traffic limit of an appropriate normalization of the storage occupancy process. In the second case, a weighted sum of Ornstein-Uhlenbeck processes describes the heavy traffic limit.

In the single-server case the results are extended to systems imposing a bound on the number of items present at any time. The limit process is Brownian motion with two reflecting barriers, a process that is conjectured to be also the appropriate diffusion limit when the bound is placed on the total storage occupancy, rather than the number in the system. The limit theorems are supplemented by discussions of the nature of the approximation, extensions to more general systems, approximations for stable systems, and the calculation of performance measures such as first passage times and stationary distributions.

3. Stochastic Models of Two-Level Memories

Both physical and economical constraints place limits on the capacity of random access primary storage devices. Since these limits are often exceeded by the storage demands of executing programs, primary storage is supplemented by auxiliary or secondary storage devices such as magnetic disks or drums. Such devices provide larger capacities at smaller unit costs. The potential sacrifices are the slower speed of non-random access devices and the delays required by indirect accesses; i.e. whenever information in secondary but not main memory is needed by an executing program it must first be transferred to main memory. Stochastic analyses of non-random access secondary storage devices are covered in chapter 6.

It is obviously desirable to include as one of the functions of an operating system the control of the interaction between primary and secondary storage. In this way the storage hierarchy is transparent to the user; he constructs programs under the assumption that the union of primary and secondary storage is available during execution of his program. This logical organization of computer memory is known as virtual memory and is a common feature of general purpose systems.

In the implementation of virtual memory a number of important problems arise. First, efficient mechanisms must be designed to map the addresses specified by a program into the actual addresses in physical memory where the corresponding information is currently stored. The details of the hardware and software structures for address mapping are not material to our purposes, and are therefore not within the scope of this book. The interested reader will find introductory treatments of this aspect of computer design in any basic text on computer architecture or operating systems.

The fragmentation problems discussed in the preceding section must also be resolved. Garbage collection and the process of maintaining and searching available space lists can require significant amounts of time and space.

Paging systems were a major development in system design for dealing with the two problems we have mentioned. In such systems program and data files are partitioned into equal-size blocks, called pages, the last of which may only be partially occupied. The page then becomes the unit of information storage and transfer. The fitting problems illustrated in the preceding section are eliminated, since all holes in storage must accommodate at least one page. Also, control structures are greatly simplified by not having to maintain the lengths of blocks of information.

Clearly, these advantages are offset by the fact that a page of information is not generally an inherent storage unit of the algorithms implemented by programs. However, by judicious selection of page size (e.g. see [GelTB,Hat]) and the other parameters introduced below, the corresponding adverse effects can be more than compensated by the advantages of such systems.

In chapters 4 and 5 we confine our study of two-level memories to paging systems. Fortunately, the practical advantages of paging systems are accompanied by major simplifications in the mathematical models; effectively, the assumption of equal size pages provides a critical reduction in the dimensionality of the problem. This is illustrated also in section 3.6 on dynamic storage allocation.

Acceptable performance of a two-level memory system depends on the ability to maintain the number of input/output operations involving relatively slow secondary storage devices at a reasonably low level. Thus, having chosen a paging system organization, we now have the problem of designing algorithms for deciding at each point in time what subset of a program's pages is to be kept in primary memory. Specifically, the central problem studied in chapters 4 and 5 is the sequencing of page loading and replacement so as to minimize, at least approximately, the number of instances in which pages not in main memory are referenced by an executing program.

To define this problem in yet finer terms we need to specify modes of operation. That mode receiving the most attention in chapter 4 is termed *demand pag-*

ing. In this mode the operating system begins by installing some initial subset of a program's pages in the main memory. Thereafter, the subset of pages in main memory is changed only when the program references a page not in main memory. In this event, called a *page fault*, the system transfers the referenced page from a secondary storage device to main memory. However, if main memory is full at the time of the page fault, it may be necessary to precede the above operation by a transfer of some page from main memory to secondary storage, so as to make room for the new page. The rule for deciding which page to remove from main memory is embodied in what is called the *replacement algorithm*. As reflected in chapter 4 the design and analysis of replacement algorithms is at the heart of performance studies of paging systems.

Our stochastic models entail a number of simplifications commonly found in the literature. Examples of the effects not explicitly taken into account are:

- Selecting a page to be removed from main memory may be influenced by whether it has been changed since it was loaded. In many systems copies of all pages are retained in secondary memory; thus, pages which have not changed since being copied into main memory need not be physically returned to secondary storage.
- That subset of a program which is chosen for initial loading will have an effect on subsequent paging behavior. We concentrate on those situations where this effect is only transient, and adopt simplistic assumptions about the start-up procedures for program execution.

A characteristic assumption underlying the design and analysis of replacement algorithms is that the sequence of pages referenced by an executing program is not known in advance. Thus, the design of replacement rules is essentially a problem in adaptive control; replacement decisions are based on probability statements about the future referencing behavior, which in turn are based on data collected on past referencing behavior.

In a typical problem formulation the dynamic storage demand of a program is represented by a page reference string, e.g. a string of integers indexing consecutively the pages referenced. In a stochastic model, a probability law governing successive references is given. For example, in the *independent reference model* the indices of the pages referenced are assumed to be independent, identically distributed random variables. A focal point of the analysis is then the process $\{X_i, i \geq 0\}$, defined as the subset of the program's pages in main memory just after the ith page reference.

A trajectory of the process $\{X_i\}$ is obtained as a function of a given page reference string; this function is defined by the page replacement rule and has the main memory size as a parameter. A simple but important example is provided by the FIFO (first-in-first-out) replacement rule: when main memory is full, the next page to be replaced is that one having been in memory for the longest time. In general, of course, $\{X_i\}$ is not a Markov chain even when the page reference string is. However, under the independent-reference model of the FIFO rule for example, we can obtain a Markov chain by imposing an ordering on the elements of $\{X_i\}$ which reflects the FIFO criterion.

In chapters 4 and 5 this model is analyzed along with a number of others differing in the replacement rule and the probabilistic assumptions about page referencing behavior. We also study an optimal algorithm for page replacement; since the information it requires is not normally available, the value of this algorithm lies chiefly in our ability to test the relative performance of approximation rules like FIFO.

So far we have considered the operation of a single program in a paged two-level memory. However, it is also important to study multiprogramming or multiprocessing systems in which a number of programs share a common two-level memory system. An initial question posed by the design of such systems is how to partition main memory among the executing programs. The most easily implemented approach is to divide up memory equally among these programs and to force each to operate wholly within its fixed memory allotment. The paging system analysis referred to earlier thus applies to each program individually.

However, such a system does not exploit the dynamic nature of a typical program's memory requirement; over time the amount of main memory required for the efficient operation of a program may vary substantially. To make use of this behavior paging policies have been studied which attempt to retain in memory only those pages of each program necessary to keep the program's rate of page faults at an acceptably low level.

Such subsets of pages have been called *working-sets*. The corresponding policies attempt to capture the changing *locality* exhibited by page referencing patterns. Roughly speaking, the locality property applies to programs whose page references are confined to relatively small subsets of pages that change slowly with time.

A policy that has received a great deal of study [Den3] defines this working set at any given time as the subset of pages that have been referenced in the preceding τ references, where τ is a parameter that can be adjusted to vary the page fault rate. This working-set policy then allows execution of a program only if its entire working set is in main memory. Thus, programs with temporarily large working sets can be accommodated effectively with programs having temporarily small working sets. When the total demand of current working sets exceeds the memory available, one or more programs can be de-activated; conversely, whenever sufficient unneeded space accumulates, additional working sets can be established in main memory. A stochastic analysis of working sets is a major thrust of chapters 4 and 5.

In chapter 4 basic definitions are presented and results mainly of a theoretical nature are derived with an emphasis on the independent reference model. Pragmatic performance evaluation issues are covered in chapter 5; approximations and efficient calculations based on a realistic model of program behavior are featured.

The material of chapter 4 is based on the work of Aven and Kogan which was first published in Russian [AveK3]. In order to keep our treatment to a reasonable size we have had to exclude some important topics, a number of which are reviewed at the end of chapter 4. However, the literature on the subject of chapters 4 and 5 is indeed large (see [Smi3] for an extensive bibliography). The following surveys are recommended to the interested reader. In [Den2, Smi4] the details of hardware and software structures for two-level memory management and the techniques for their analysis are reviewed. The history of memory management at IBM during the period extending from the late 1950's to the early 70's is discussed in [BelaPS]. The working set concept and its important implications for optimal multiprogramming control as well as a critical review of program behavior models can be found in [Den3].

4. Stochastic Models of Secondary Storage

As illustrated in the previous section, important congestion points in computer systems frequently occur through interactions with secondary storage devices. Thus, overall system performance often depends substantially on the techniques used to handle the requirements for these devices. Chapter 6 presents detailed analyses of the performance of the more important secondary storage devices. The dependence of performance on the physical characteristics and methods of operation is also studied in some detail.

The analysis concentrates on drum and disk-like devices because of their general importance; however, many of the basic results apply to other devices such as tapes, magnetic bubble memories, etc. As pointed out earlier the reader is assumed to be familiar with the basic organization of disk and drum systems. Knowledge of design details, especially those not common to all such systems, is not required to follow our treatment.

The dynamics of moving-arm and rotating-storage devices determine the optimization techniques that can be applied to their use. Drums are intrinsically simpler than disks, for only one variable is needed to describe the state of the device; the angular position of the drum as a function of time completely describes the physical configuration of the device. With disk devices, on the other hand, two state descriptors are required: the angular position of the disk and the (radial) position of the arm. Thus, the latency optimization problem resulting from the angular displacements of addressed locations is common to the two devices.

Let us now consider scheduling policies dealing with the two sources of delay.

Arm Positioning (Seek) Delays - The simplest policy to implement for scheduling a disk arm, in terms of the required data structures and hardware, is a linear FIFO queue, and it is this policy that is analyzed in section 6.2. It is easily shown and intuitively clear that FIFO sequencing results in many more, and relatively longer seeks than alternatives require. Any improvement requires keeping track of the requests according to their target cylinders. Because of the relatively long time required to perform a seek, most of the improved policies schedule all of the requests for the cylinder currently under the read/write heads before all other pending requests.

There are two such policies that have been adopted for the solution of the arm scheduling problem. The first is SSTF, an acronym for shortest-seek-time-first. Under this policy, after all the requests to the current cylinder are served, the arm performs a seek to the nearest cylinder that has a waiting request. Under certain conditions this policy is thought to achieve the shortest over-all mean waiting time, although no successful analysis has yet appeared. At the same time, however, the variance of this time may be undesirably large when the input of requests is inhomogeneous in time. These issues are further discussed in [Hof1].

The second is SCAN. Under this policy the motion of the arm is organized so as to reduce the variance of the waiting time when compared to SSTF. The sacrifice made for this improvement is an increase in the mean waiting time. More specifically, at any given time instant the arm is in one of two modes, 'in', or 'out'. When a seek is required in the first mode, the arm moves toward the disk spindle until a cylinder with pending requests is encountered. When no such cylinder exists, the mode of the arm is changed on reaching the boundary and its direction of motion reversed. Arm movement in the out mode is similar; thus the arm oscillates over the disk. This sequencing rule is analyzed in section 6.3 following the work of Coffman and Hofri [CofH2].

Two relatively simple and less accurate models of the SCAN policy were studied in [GotM] and [One]. In [AveGK] a diffusion approximation with variable drift and diffusion coefficient was developed for these models and its accuracy evaluated.

Rotational Latency Delays - FIFO sequencing of requests for devices imposing delays in rotation is embraced in the general model of section 6.2. The principal technique for reducing latency effects is called SLF (smallest-latency-first) scheduling, and can result in improvements in capacity by a factor of N, the number of records accessible in one drum revolution. At each decision point the next request served is that one of those waiting whose starting address is nearest to the current angular position of the drum. Although a general analysis of SLF systems is difficult, in the important context of paging systems results are rather easily obtained through a Markov chain analysis. In section 6.2 we show how these results can be obtained from the results on FIFO systems.

Multiple Head Systems - In section 6.4 we analyze a mathematical model of computer disk storage devices having two movable read/write heads. The model was first introduced in [PagW]; our analysis closely follows that in [CalCF2]. The model approximates the set of storage addresses by the continuous interval [0,1]. Arriving requests for information at points on this interval are assumed to be processed in strict first-come-first-serve order. These requests are modeled as a sequence of independent random variables uniformly distributed over [0,1]. Processing a request consists first of selecting the head to process the request, then positioning this head at the requested location, and finally, carrying out the read/write operation itself.

In [CalCF1] a number of head selection policies were examined within this model, each assuming that the two heads could be positioned independently. In particular, if the head positions after processing the ith request were $0 \leqslant x \leqslant y \leqslant 1$, and the $(i+1)^{st}$ request was at point z, then for all head-selection rules the new head position was (z,y) if $z \leqslant x$, and (x,z) if $z \leqslant y$. But either (z,y) or (x,z) could be chosen when $x \leqslant z \leqslant y$, so it was the criterion for this choice that distinguished different policies. Subsequently, a variant of this model was analyzed [Hof2] in which the head not selected to process a request was allowed to move as well. It was assumed that this latter motion could always be performed within the time interval required by the other head to process the request.

In section 6.4 we analyze the two-head system under the assumption that the two heads must remain locked in a fixed relative position, a distance d apart. Thus, when servicing a request both heads must move: One is moved to the request location, and the other is moved simultaneously an equal distance in the same direction. Both heads must always remain within [0,1]. Therefore, coverage of the entire interval by the two heads requires that $0 \leqslant d \leqslant 1/2$. Moreover, it is clear that the left-head is restricted to the interval $[0, 1-d]$ while the right-head is restricted to $[d, 1]$.

As in the other studies the major objective is the expected distance that the heads must move in processing a request under the so-called *nearer-server* rule: For requests in $[d, 1-d]$ the nearer of the two heads is always selected to process the request. (As implied above, requests in $[0, d]$ and $[1-d, 1]$ must always be processed by the left and right head respectively.) This performance measure is calculated as a function of d, and then its minimum over $0 \leqslant d \leqslant 1/2$ is found.

Relative to other proposed two-head systems the one investigated in section 6.4 is easily motivated, for it entails only a single positioning mechanism. Such systems are markedly less expensive to build than those modeled earlier, where two independent head positioning mechanisms must be implemented. References to specific systems maintaining two heads a fixed distance apart can be found in [PagW]. Simulation studies under less restrictive assumptions are also presented there.

Chapter 2

Buffer Storage

1. Introduction

The streams of messages generated at terminals or of data generated by computers are quite often sporadic. In order to enhance the utilization of communication links, a technique such as buffering is commonly adopted. A buffer store is interposed between the set of incoming links and an output link; its general purpose is to receive and store information according to a stochastic process defining the source, and then to output the information according to a different process that defines the sink. In this chapter we shall restrict ourselves to an output process which is simply a smoothing of the input process; i.e. the output of a non-empty buffer is at a constant rate which is normally assumed to be less than the source rate. A central design parameter is the buffer size, which must be sufficiently large to accommodate large backlogs of messages or data units that may occasionally develop because of large, transient demands of the input process. Accordingly, we shall discuss the problem of finding the steady state distribution of the amount of information stored in the buffer.

The unit of information flow through a buffer system is called an item. Arriving items are characterized by a statistical law governing their sizes, which corresponds to the service time distribution in the usual queueing system. A major source of variation among buffer systems is the *loading protocol*, i.e. the mechanism describing how arriving items enter the buffer. In general, it is necessary to assume that it takes time to load or transmit an item, a process that typically requires a time interval proportional to the item's size or length. Thus, the size of arriving items will determine both the time it takes to load them and the space they occupy in the buffer. This assumption along with a constant output rate are characteristic of the models analyzed in the next two sections. For a discussion of other possibilities and the related literature see section 1.2.

In defining the loading protocol it is necessary to specify whether "flow-through" is implemented. Flow-through refers to the process whereby the output

18

of an arriving item begins the moment that it has nothing ahead of it, as opposed to having to wait until the entire item has been loaded. Thus, if the capacity is c, the loading rate is $r > 1$ and the output rate is $r_o = 1$, then an item finding x in the buffer can begin output after a delay of x time units if its size z is at most $(c-x) \cdot r/(r-1)$, for in that case the buffer will not overflow.

In contrast, there are systems in which it is necessary to load an entire item into a buffer register first, and then to transfer it all at once (in parallel) into the store. For the same parameters as before there is enough room in the buffer at the end of loading provided that

$$z \leqslant c - (x - z/r)^+,$$

where $y^+ = \max(0, y)$ for any real y.

We study both types of systems in sections 2 and 3. In yet another type of system items are loaded directly into the buffer during transmission, but output does not occur until after loading has terminated. Although we do not consider such systems, the methods we use apply to their analysis as well (see [GavM,GavL]).

Finally, we must specify how the system treats arriving items for which there is not enough room in the buffer. There are several possibilities for handling overflow items: (1) The buffer could attempt to accept all items, but stop loading whenever the buffer content reaches c; at such times the remainder of an item currently being loaded would be lost. (2) The buffer could operate as above, but in the case of overflow the content would be held at c for the time it would take to load the remainder of the current item. (3) Assuming that the sizes of arriving items are known (given in "headers", for example), the buffer could accept only items that would not overflow the store. The first and third of these assumptions are referred to as "reflection" and "non-overflowing" protocols, respectively, and are examined in detail in what follows.

It is not difficult to conceive of other possibilities for loading and overflow management, each creating a different model. We consider only a few of the more representative models in order to suggest to the reader how to work out others by similar methods. The analysis of the models that we have chosen follows closely the recent work of Beneš [Ben1].

2. Equilibrium Equations for Models with a Single Source

In this section we assume that the sizes of arriving items are i.i.d. random variables with the general distribution function B, and that the intervals between loading completions and the earliest subsequent arrival times are independently and exponentially distributed with mean $1/\lambda$. For convenience we adopt the normalization $r_o = 1$ in this section and the next. Finally, to be consistent with the need for storage in systems with flow-through, we assume $r > r_o = 1$. For other systems we also consider the case $r = 1$.

Loading with flow-through and infinite capacity - We assume that an arriving item can be loaded directly into the store, and that it takes z/r time units to load an item of size z. An item can begin output as soon as any part of it has been loaded.

The system will have three operating modes: an emptying mode when nothing is being loaded and the buffer content is decreasing at rate -1; an emptying-loading mode during which an arrival is being loaded at rate $r > 1$ while output is taking place; and a totally-empty mode with nothing stored and no loading taking place. Because of $r > 1$ and the flow-through assumption, the buffer content actually grows at rate $r - 1$ during the emptying-loading mode. In this subsection we assume an infinite capacity, $c = \infty$. The finite-capacity case is studied in the next subsection.

Let x_t be the store content at time t, and let τ_t be the amount of time spent prior to t in loading the item currently entering the store. It can be seen from our assumptions that the process whose values are defined as

$$
\begin{array}{ll}
(0, x_t) & \text{in the emptying mode,} \\
(\tau_t, x_t) & \text{in the emptying-loading mode,} \\
(0, 0) & \text{when the system is empty,}
\end{array}
$$

is a Markov process.

To formulate forward equations for the process (τ_t, x_t) we first note that $\tau_t \leqslant x_t/(r-1)$ a.s. (almost surely) since $r > 1$, i.e. the time already spent loading an item cannot exceed the buffer content divided by $(r-1)$, and equality occurs if and only if the item found the buffer empty on arrival. Indeed, an arriving item of size z finding the buffer empty will take z/r time units to be loaded; during this time τ_t will increase linearly at rate 1, x_t will increase linearly at rate $r-1$, and (τ_t, x_t) will move along the line $\tau = x/(r-1)$ for a time z/r. With a positive probability of finding an empty buffer, the line $\tau = x/(r-1)$ will have positive probability, and the joint distribution of (τ_t, x_t) will have jumps across this line. Thus, the emptying-loading mode consists of two sub-modes, one corresponding to motion along $\tau = x/(r-1)$ due to finding an empty buffer, and the other to motion along lines $\tau - x/(r-1) = $ constant < 0.

Thus the Markov process (τ_t, x_t) travels along straight lines while in a mode; it changes mode at rate λ when not loading, and at rate $r\, a\, (r\tau)$ when $\tau_t = \tau$, where

$$a(y) = \frac{b(y)}{1 - B(y)} \, , \quad b(y) = B'(y),$$

is the failure rate associated with the distribution B of item sizes.

We use the following notation for state probabilities:

$$Q(x,t) = Pr\{x_t \leqslant x, \text{ no loading}\},$$
$$P(\tau,x,t) = Pr\{\tau_t \leqslant \tau, x_t \leqslant x, (r-1)\tau_t < x_t\},$$
$$N(\tau,t) = Pr\{\tau_t \leqslant \tau, x_t = (r-1)\tau_t\}.$$

The respective modes of operation will be called the Q, P and N modes.

To relate these probabilities we consider the possible events in a small time interval $[t, t+\Delta t]$. We can reach the state $\{x_{t+\Delta t} \leqslant x, \text{ no loading}\}$

(i)   if the state at time t was $\{x_t \leqslant x+\Delta t, \text{ no loading}\}$ and there were no arrivals during $[t, t+\Delta t]$ (which occurs with probability $1 - \lambda\Delta t + o(\Delta t)$), or

(ii)   if the state at time t was $\{u < \tau_t \leqslant u+du, z < x_t \leqslant z+dz, (r-1)\tau_t < x_t\}$ for some $0 \leqslant u \leqslant z/(r-1)$ and $0 \leqslant z \leqslant x$, and loading was completed in $[t, t+\Delta t]$ (which occurs with probability $r\, a\, (ru)\Delta t + o(\Delta t)$), or

(iii)   if the state at time t was $\{u < \tau_t \leqslant u+du, x_t = (r-1)\tau_t\}$ for some $0 \leqslant u \leqslant x/(r-1)$ and loading was completed in $[t, t+\Delta t]$ (which occurs with probability $ra(ru)\Delta t + o(\Delta t))$. Since all other events during $[t, t+\Delta t]$ have probabilities $o(\Delta t)$, the above observations lead to

$$Q(x, t+\Delta t) = [1-\lambda\Delta t]Q(x+\Delta t, t) + r\Delta t \int_0^{x} \int_0^{z/(r-1)} a(ru)P(du\,dz, t)$$

(1)
$$+ r\Delta t \int_0^{x/(r-1)} a(ru)N(du, t) + o(\Delta t), \quad x \geqslant 0.$$

Expanding $Q(x+\Delta t, t)$ with respect to its first variable we have

$$Q(x+\Delta t, t) = Q(x, t) + \Delta t \frac{\partial Q(x, t)}{\partial x} + o(\Delta t).$$

Using this expansion for the first term on the righthand side of (1) produces

$$Q(x, t+\Delta t) = Q(x, t) + \Delta t \frac{\partial Q(x, t)}{\partial x} - \lambda\Delta t Q(x, t)$$
$$+ r\Delta t \int_0^{x} \int_0^{z/(r-1)} a(ru)P(du\,dz, t) + r\Delta t \int_0^{x/(r-1)} a(ru)N(du, t) + o(\Delta t).$$

Thus, ignoring terms which are $o(\Delta t)$ we obtain

$$\frac{Q(x, t+\Delta t) - Q(x, t)}{\Delta t} = \frac{\partial Q(x, t)}{\partial x} - \lambda Q(x, t) +$$
$$+ r \int_0^{x} \int_0^{z/(r-1)} a(ru)P(du\,dz, t) + r \int_0^{x/(r-1)} a(ru)N(du, t).$$

Passage to the limit $\Delta t \to 0$ yields the Q-equation

(2)
$$\frac{\partial Q(x, t)}{\partial t} = \frac{\partial Q(x, t)}{\partial x} - \lambda Q(x, t) + r \int_0^{x} \int_0^{z/(r-1)} a(ru)P(du\,dz, t)$$
$$+ r \int_0^{x/(r-1)} a(ru)N(du, t).$$

By similar arguments we derive

$$P(\tau, x, t+\Delta t) = \int_0^{x-(r-1)\Delta t} \int_0^{\tau-\Delta t} (1 - r\Delta t a(ru))P(du\,dz, t)$$

(3)
$$+ \lambda\Delta t Pr\{0 < x_t \leqslant x, \text{ no loading}\} + o(\Delta t)$$
$$= P(\tau-\Delta t, x-(r-1)\Delta t, t) - r\Delta t \int_0^{x-(r-1)\Delta t} \int_0^{\tau-\Delta t} a(ru)P(du\,dz, t)$$

$$+ \lambda \Delta t [Q(x,t) - Q(0,t)] + o(\Delta t), \tau > 0, x > 0,$$

and

$$N(\tau, t + \Delta t) = \int_0^{\tau - \Delta t} (1 - r\Delta t a(ru)) N(du,t) + \lambda \Delta t Pr\{x_t = 0, \text{no loading}\} + o(\Delta t),$$

$$(4) \qquad = N(\tau - \Delta t, t) - r\Delta t \int_0^{\tau - \Delta t} a(ru) N(du,t) + \lambda \Delta t Q(0,t) + o(\Delta t), \tau > 0.$$

Expanding $P(\tau - \Delta t, x - (r-1)\Delta t, t)$, with respect to its first and second variables we have

$$(5) \qquad P(\tau - \Delta t, x - (r-1)\Delta t, t) = P(\tau, x, t) - \Delta t \frac{\partial P(\tau, x, t)}{\partial \tau}$$

$$+ (1-r)\Delta t \frac{\partial P(\tau, x, t)}{\partial x} + o(\Delta t).$$

Using this expansion for the first term on the righthand side of (3) we obtain routinely as $\Delta t \to 0$ the P-equation

$$\frac{\partial P(\tau, x, t)}{\partial t} = (1-r) \frac{\partial P(\tau, x, t)}{\partial x} - \frac{\partial P(\tau, x, t)}{\partial \tau}$$

$$- r \int_0^x \int_0^\tau a(ru) P(du\, dz, t) + \lambda [Q(x,t) - Q(0,t)].$$

Similarly, for the N-equation we conclude that

$$(6) \qquad \frac{\partial N(\tau, t)}{\partial t} = - \frac{\partial N(\tau, t)}{\partial \tau} - r \int_0^\tau a(ru) N(du,t) + \lambda Q(0,t).$$

Assuming the existence of the limits

$$Q(x) = \lim_{t \to \infty} Q(x,t), \quad P(\tau, x) = \lim_{t \to \infty} P(\tau, x, t), \quad N(\tau) = \lim_{t \to \infty} N(\tau, t),$$

we have from (2), (5) and (6) the equilibrium equations

$$(7) \qquad 0 = \frac{dQ}{dx} - \lambda Q + r \int_0^x \int_0^{z/(r-1)} a(ru) P(du\, dz) + r \int_0^{x/(r-1)} a(ru) N(du)$$

$$(8) \qquad 0 = (1-r) \frac{\partial P}{\partial x} - \frac{\partial P}{\partial \tau} - r \int_0^x \int_0^\tau a(ru) P(du\, dz) + \lambda [Q(x) - Q(0)]$$

(9) $0 = -\dfrac{dN}{d\tau} - r\displaystyle\int_0^\tau a\,(ru)N\,(du) + \lambda Q\,(0),$

along with the normalization condition that the probabilities sum to 1.

The stationary equations are more easily understood if we introduce the following reasonable assumptions:

(i) Q has an atom $Q\,(0)$ at the origin, corresponding to a positive probability that the system is empty, but is otherwise absolutely continuous with a smooth density q.

(ii) P and N have respective smooth densities p and n ($N(y)$ is the stationary probability assigned to the segment $\tau = x/(r-1)$, $0 \leqslant \tau \leqslant y$, in the (τ, x)-plane as a result of arrivals finding an empty buffer).

These assumptions can be seen to apply to the solutions derived in the following section. We can now differentiate (7) with respect to x and find that

(10) $0 = \dfrac{dq}{dx} - \lambda q + r\displaystyle\int_0^{x/(r-1)} a\,(r\dot u)p\,(u,x)\,du + a\left(\dfrac{rx}{r-1}\right)n\left(\dfrac{x}{r-1}\right)\dfrac{r}{r-1}.$

Applying $\dfrac{\partial^2}{\partial\tau\partial x}$ to (8) produces

(11) $\dfrac{\partial p}{\partial\tau} = (1-r)\dfrac{\partial p}{\partial x} - ra\,(r\tau)p,$

and applying $-\dfrac{\partial}{\partial\tau}$ to (9) yields

(12) $0 = -\dfrac{dn}{d\tau} - r\,a\,(r\tau)n.$

Equation (1) at $x = 0$ implies

(13) $q\,(0) = \lambda Q\,(0).$

Boundary conditions for p and n are implied by (3) and (4) respectively at $\tau = \Delta t$. Specifically,

(14) $p\,(0,x) = \lambda q\,(x),$
(15) $n\,(0) = \lambda Q\,(0).$

Flow-through and finite capacity, reflection protocol - When x_t reaches c in either the P or the N mode loading ceases, the rest of the item is lost and the system moves directly to the Q mode. The equations remain as above, but they apply only for $0 \leqslant x \leqslant c$. The reflection at $x = c$ is expressed by the boundary condition

(16)
$$q(c) = (r-1) \int_0^{c/(r-1)} p(u,c)\,du + n(c/(r-1)),$$

which incorporates the fact that the reflecting trajectories of x_t arrive at c with slope $r-1$ and leave with slope -1. To derive (16) we observe that for $\Delta t > 0$ sufficiently small, the event $x_{t+\Delta t} > c - \Delta t$ in the Q mode is preceded by the event $x_t > c - (r-1)\Delta t$ in the P mode, given that loading is not completed in the interval $(t, t+\Delta t)$, or the event $\tau_t > c/(r-1) - \Delta t$ in the N mode, given that loading is not completed in the interval $(t, t + \Delta t)$. These observations lead us to

$$\int_{c-\Delta t}^c Q(dz, t+\Delta t) = \int_{c-(r-1)\Delta t}^c \int_0^{z/(r-1)} [1 - \Delta t\, ra(ru)] P(du\, dz, t)$$
$$+ \int_{c/(r-1)-\Delta t}^{c/(r-1)} N(du, t) \left[1 - \Delta t\, ra\left(\frac{c}{r-1} - \Delta t\right)\right] + o(\Delta t).$$

Passing to the limit $t \to \infty$ and using the fact that the stationary distributions Q, P and N have smooth densities near the boundary c we obtain

$$q(c)\Delta t = (r-1)\Delta t \int_0^{c/(r-1)} p(u,c)\,du + \Delta t\, n\left(\frac{c}{r-1}\right) + o(\Delta t).$$

Dividing by Δt and passing to the limit $\Delta t \to 0$ produces (16).

Flow-through and finite capacity, non-overflowing protocol - When the size of an item is known at its time of arrival (before it starts loading), it is possible in principle to accept only those items that will not cause overflow beyond the capacity c. If an item of size z arrives with x in the buffer, it will take z/r time units to load and hence it will not cause an overflow if and only if

$$z + x - z/r \leqslant c.$$

Let us use the amount x_t in the buffer and the amount z_t left to be loaded as the basic random processes, with probabilities $Q(x,t) = Pr\{x_t \leqslant x$, no loading$\}$

when no loading is in progress, and $P(x,z,t) = Pr\{x_t \leqslant x, z_t \leqslant z\}$ when loading is taking place. To derive equilibrium equations we introduce the random event

$$A_t = \{\text{no arrival occurs in the interval } (t,t + \Delta t)\}$$

and its complement $\overline{A}_t$. Denote by ξ the random size of an arriving item. As before we wish to relate the probabilities $Q(x,t + \Delta t)$ and $P(x,z,t + \Delta t)$, where $\Delta t > 0$ is small. For $Q(x,t + \Delta t)$ we observe that the event $\{x_{t+\Delta t} \leqslant x, \text{no loading}\}$ is preceded by intersections of the events $\{x_t \leqslant x + \Delta t, \text{no loading}\}$ and $\overline{A}_t$, or by intersections of the events A_t and $\{$for some $y \in [0,x + \Delta t], y < x_t \leqslant y + dy, \xi > r(c-y)/(r-1)\}$, or by the event $\{x_t \leqslant x - (r-1)\Delta t , 0 < z_t \leqslant r\Delta t\}$. These observations yield

$$(17) Q(x,t + \Delta t) = (1 - \lambda\Delta t) Q(x + \Delta t,t)$$
$$+ \lambda\Delta t \int_0^{x+\Delta t} \left[1 - B\left[(c-y)r/(r-1)\right]\right] Q(dy,t)$$
$$+ P(x - (r-1)\Delta t, r\Delta t,t) - P(x - (r-1)\Delta t,0,t) + o(\Delta t).$$

Subtracting $Q(x,t)$ from both sides of this equation, dividing by Δt, and passing to the limit $\Delta t \to 0$, we finally obtain the first differential equation

$$(18) \qquad \frac{\partial Q(x,t)}{\partial t} = \frac{\partial Q(x,t)}{\partial x} - \lambda \int_0^x B\left[\frac{(c-y)r}{r-1}\right] Q(dy,t)$$
$$+ rP'_z(x,0,t), 0 < x \leqslant c ,$$

where $P'_z(x,0,t) = \frac{\partial P(x,z,t)}{\partial z}\big|_{z=0}$. By similar observations we conclude that

$$(19) \qquad P(x,z,t + \Delta t) = P(x-(r-1)\Delta t, z+r\Delta t,t)$$
$$+ \lambda\Delta t Q(x+\Delta t,t) B(z) + o(\Delta t),$$

and hence the second differential equation

$$(20)\ \frac{\partial P(x,z,t)}{\partial t} = (1-r)\frac{\partial P(x,z,t)}{\partial x} + r\frac{\partial P(x,z,t)}{\partial z} + \lambda Q(x,t)B(z),$$
$$0 < [(r-1)/r]z \leqslant c-x .$$

Assuming that the stationary probability distributions

$$Q(x) = \lim_{t \to \infty} Q(x,t) \text{ and } P(x,z) = \lim_{t \to \infty} P(x,z,t)$$

have the same properties as before we obtain from (18) and (20) the following equilibrium equations for the respective densities $q(x)$ and $p(x,z)$.

(21) $$0 = \frac{dq(x)}{dx} - \lambda B \left[\frac{r(c-x)}{r-1} \right] q(x) + rp(x,0), \; 0 < x \leqslant c,$$

(22) $$0 = (1-r)\frac{\partial p(x,z)}{\partial x} + r\frac{\partial p(x,z)}{\partial z} + \lambda q(x)b(z), \; 0 < z \leqslant (c-x)r/(r-1) .$$

The boundary conditions

(23) $$q(0) = \lambda Q(0)$$

and

(24) $$(r-1)p(0,z) = \lambda Q(0)b(z)$$

are implied by (17) at $x = 0$ and (19) at $x = (r-1)\Delta t$, respectively. (From our assumptions we note that $P(0,z,t) \equiv 0$).

Loading followed by transfer - We consider next the same model as before, except that flow-through does not occur; arriving items must be loaded at rate $r \geqslant 1$ into a register, and then transferred into the buffer in parallel. During loading of the register the buffer content drops at rate -1 as before until it reaches and then remains temporarily at zero, or loading ends and the new item is added to the buffer.

The process describing buffer operation will have three modes: a simple emptying mode described by an amount x_t in the buffer; a loading mode with the buffer empty (prior to transfer), described by an amount y_t loaded so far and an amount z_t remaining; and a simultaneous loading-emptying mode described by all of x_t, y_t, and z_t. We let $Q(x,t), N(y,z,t)$, and $P(x,y,z,t)$ be the probability distributions corresponding to the three modes in the order given.

We first consider the case of infinite capacity. We have

$$P(x,y,z,t + \Delta t) = Pr\{x_{t+\Delta t} \leqslant x, y_{t+\Delta t} \leqslant y, z_{t+\Delta t} \leqslant z \text{ in the } P \text{ mode}\}$$
$$= Pr\{x_t \leqslant x+\Delta t, y_t \leqslant y-r\Delta t, z_t \leqslant z+r\Delta t \text{ in the } P \text{ mode}\}$$
$$= P(x+\Delta t, y-r\Delta t, z+r\Delta t, t), \; x > 0, \; y > 0, \; z > 0.$$

Subtracting $P(x,y,z,t)$ from both sides of this equation, dividing by Δt and passing to the limit $\Delta t \rightarrow 0$, we arrive at the first differential equation

$$(25) \quad \frac{\partial P(x,y,z,t)}{\partial t} = \frac{\partial P(x,y,z,t)}{\partial x} - r\frac{\partial P(x,y,z,t)}{\partial y} + r\frac{\partial P(x,y,z,t)}{\partial z}.$$

The relation

$$N(y,z,t + \Delta t) = Pr\{y_{t+\Delta t} \leqslant y, z_{t+\Delta t} \leqslant z \text{ in the } N \text{ mode}\}$$
$$= Pr\{y_t \leqslant y - r\Delta t, z_t \leqslant z + r\Delta t \text{ in the } N \text{ mode}\}$$
$$+ Pr\{0 < x_t \leqslant \Delta t, y_t \leqslant y - r\Delta t, z_t \leqslant z + r\Delta t \text{ in the } P \text{ mode}\} ;$$
$$y > 0, z > 0$$

leads us to the second differential equation

$$(26) \quad \frac{\partial N(y,z,t)}{\partial t} = -r\frac{\partial N(y,z,t)}{\partial y} + \frac{\partial N(y,z,t)}{\partial z} + P'_x(0,y,z,t),$$

where

$$P'_x(0,y,z,t) = \frac{\partial P(x,y,z,t)}{\partial x}\Big|_{x=0}.$$

The differential equation that we need for Q is implied by

$$(27)\ Q(x,t+\Delta t) = Pr\{x_{t+\Delta t} \leqslant x \text{ in the } Q \text{ mode}\} =$$
$$= Pr\{x_t \leqslant x+\Delta t \text{ in the } Q \text{ mode and } \overline{A}_t\}$$
$$+ Pr\{x_t+y_t \leqslant x+\Delta t-r\Delta t, 0 < z_t \leqslant r\Delta t \text{ in the } P \text{ mode}\}$$
$$+ Pr\{y_t \leqslant x-r\Delta t, 0 < z_t \leqslant r\Delta t \text{ in the } N \text{ mode}\} =$$

$$= [1 - \lambda\Delta t + o(\Delta t)]Q(x+\Delta t,t) + N(x-r\Delta t, r\Delta t,t)$$
$$- N(x-r\Delta t,0) \quad + \iint_{0 \leqslant u+v \leqslant x-(r-1)\Delta t} [P(du\,dv,r\Delta t) - P(du\,dv,0)].$$

This yields

$$\frac{\partial Q(x,t)}{\partial t} = \frac{\partial Q(x,t)}{\partial x} - \lambda Q(x,t) + r\int_0^x \int_0^{x-u} P'_z(du\,dv,0,t)$$
$$(28) \qquad\qquad\qquad + rN'_z(x,0,t) ,$$

where

$$P'_z(x,y,0,t) = \frac{\partial P(x,y,z,t)}{\partial z}\Big|_{z=0}, \quad N'_z(y,0,t) = \frac{\partial N(y,z,t)}{\partial z}\Big|_{z=0}.$$

We assume the existence of the stationary distributions

$$P(x,y,z) = \lim_{t \to \infty} P(x,y,z,t); \; N(y,z) = \lim_{t \to \infty} N(y,z,t); \; Q(x) = \lim_{t \to \infty} Q(x,t).$$

$P(x,y,z)$ and $N(y,z)$ have smooth densities $p(x,y,z)$ and $n(y,z)$, respectively, while $Q(x)$ has a jump $Q(0)$ at the origin, corresponding to a positive probability that the system is empty, and is otherwise absolutely continuous with a smooth density $q(x)$. With these assumptions we obtain from (25), (26), and (28) the following equilibrium equations for the densities:

(29)
$$0 = \frac{\partial p}{\partial x} - r\frac{\partial p}{\partial y} + r\frac{\partial p}{\partial z}$$

(30)
$$0 = -r\frac{\partial n}{\partial y} + r\frac{\partial n}{\partial z} + p(0,y,z)$$

(31)
$$0 = \frac{dq}{dx} - \lambda q + r\int_0^x p(u,x-u,0)du + rn(x,0).$$

To derive the boundary conditions we observe that

(32)
$$\begin{aligned} Pr\{x_{t+\Delta t} \leqslant x, 0 < y_{t+\Delta t} \leqslant r\Delta t, z_{t+\Delta t} \leqslant z \text{ in the } P \text{ mode}\} \\ = Pr\{x_t \leqslant x+\Delta t \text{ in the } Q \text{ mode, } A_t, \text{ and } \xi \leqslant z\} \\ = \lambda\Delta t Q(x,t)B(z) + o(\Delta t), \end{aligned}$$

(33)
$$\begin{aligned} Pr\{0 < y_{t+\Delta t} \leqslant r\Delta t, z_{t+\Delta t} \leqslant z \text{ in the } N \text{ mode}\} \\ = Pr\{x_t = 0 \text{ in the } Q \text{ mode, } A_t, \text{ and } \xi \leqslant z\} \\ = \lambda\Delta t Q(0)B(z) + o(\Delta t). \end{aligned}$$

From (32) and (33) we have

(34)
$$rp(x,0,z) = \lambda q(x)b(z)$$

and

(35)
$$rn(0,z) = \lambda Q(0)b(z),$$

respectively. The third boundary condition

(36)
$$q(0) = \lambda Q(0)$$

is implied by relation (27) at $x = 0$.

Loading with transfer, finite capacity - When the available storage space is finite, some rule for accepting and rejecting arriving items must be adopted. To define a general class of models let us suppose that there is given a function $k(x)$ to be used as follows: With x in the buffer, accept an item whose size is $k(x)$ or less, and reject all those larger than $k(x)$. Naturally, we assume that $k(x)$ is such that accepted items always fit into memory at the time of their transfer. The function k describes the overflow avoidance protocol.

By way of illustration, we consider two examples of Beneš [Ben1]. An "eager" overflow protocol accepts items for which there is not currently enough space, but for which there will be sufficient space in the buffer by the time loading is finished. For an item of size z finding x in the buffer, loading will take z/r time units, so the eager rule accepts the item if $z \leqslant c$ and $z \leqslant c - x + z/r$. In this case,

$$k(x) = \min\left\{c, \frac{r(c-x)}{r-1}\right\}.$$

On the other hand, a "lazy" protocol for avoiding overflow accepts only an item for which there is enough space at the time of arrival. For this case we define $k(x)=c-x$.

The acceptance function $k(\cdot)$ and the capacity c determine the reachable range of values of x, y and z. If y has been loaded, the item in question arrived y/r time units earlier. If x is in the buffer now, then there was $x + y/r$ at the time of arrival. Also, if z is yet to be loaded, the item size was $z + y$. Thus by the protocol at the time of arrival

$$z + y \leqslant k(x + y/r).$$

Using the notation p, n and q as previously we see that equations for p and n for the capacity and overflow protocol remain the same (i.e. (29) and (30)), but they apply only when $z + y \leqslant k(x+y/r)$ and $z+y \leqslant c$, respectively. The appropriate modification of (31) is

$$(37)\ 0 = \frac{dq}{dx} - \lambda B(k(x))q + \int\limits_{0 \leqslant y \leqslant k(x+y/r)} p(x-y,y,0)dy + rn(x,0),\ 0 < x \leqslant c.$$

The term $-\lambda B(k(x))q$ appears in the same way as the corresponding term in (21). The boundary conditions (34)-(36) also remain, but the validity of (34) and (35) is restricted to $0 \leqslant z \leqslant k(0)$ and $0 \leqslant z \leqslant k(x)$, respectively.

3. Explicit Solutions

In this section we simplify the equilibrium equations and consider certain cases where explicit solutions can be found.

Loading with flow-through - Using the definition of the failure rate $a(y)$ we rewrite (12) as

$$\frac{dn}{n} = \frac{d[1 - B(r\tau)]}{1 - B(r\tau)} ,$$

which with (15) produces

(38) $$n(\tau) = \lambda Q(0)[1 - B(r\tau)].$$

Next, we solve (11) by characteristics with initial condition (14) and obtain

(39) $$p(\tau,x) = p(0,x-(r-1)\tau)[1 - B(r\tau)]$$
$$= \lambda q(x-r\tau+\tau)[1 - B(r\tau)].$$

Substitution of (38) and (39) into (10) results in an equation with the single unknown function $q(x)$:

$$0 = \frac{dq}{dx} - \lambda q + \lambda r \int_0^{x/(r-1)} b(ru)q(x - ru+u)du + \lambda Q(0)b\left[\frac{rx}{r-1}\right]\frac{r}{r-1} .$$

We write this as

(40) $$0 = \frac{dq}{dx} - \lambda q + \lambda \int_0^x q(x-w)f(w)dw + \lambda Q(0)f(x),$$

where

$$f(w) = \frac{r}{r-1}b\left[\frac{rw}{r-1}\right]$$

is the rescaled density $b(w) = B'(w)$. This result suggests immediately that the rescaling of item sizes by $r/(r-1)$ will be a characteristic of loading with flow-through. Equation (40) is similar in form to the Takàcs equation [Tak]. Its solution can be worked out with the help of Laplace transforms. With

$$\bar{q}(s) = \int_0^\infty e^{-sx}q(x)dx \quad , \quad \bar{f}(s) = \int_0^\infty e^{-sw}f(w)dw$$

we have that

(41)
$$\bar{q}(s) = \frac{\lambda Q(0)[1 - \bar{f}(s)]}{s - \lambda[1 - \bar{f}(s)]},$$

where boundary condition (13) has been used. It can be seen that with

$$F(w) = \int_0^w f(u)du = B\left(\frac{rw}{r-1}\right),$$

$$m = \int_0^\infty [1 - F(w)]dw = \int_0^\infty uf(u)du = \frac{r-1}{r}\int_0^\infty xb(x)dx,$$

the function

$$g(w) = \frac{1 - F(w)}{m}, \quad w \geqslant 0,$$

is itself a probability density with the transform

$$\bar{g}(s) = \begin{cases} [1 - \bar{f}(s)]/sm, & s > 0, \\ \lim_{s \to 0} \{[1 - \bar{f}(s)]/sm\} = 1, & s = 0. \end{cases}$$

Thus the result for $\bar{q}$ may be rendered as

(42)
$$\bar{q}(s) = \frac{\lambda m Q(0)\bar{g}(s)}{1 - \lambda m \bar{g}(s)}.$$

To find $Q(0)$ we use the normalization

$$1 = Q(0) + \int_0^\infty q(x)dx + \int_0^\infty \int_0^{x/(r-1)} p(\tau,x)d\tau dx + \int_0^\infty n(\tau)d\tau.$$

From (42) we easily have

$$\int_0^\infty q(x)dx = \bar{q}(0) = \frac{\lambda m Q(0)}{1 - \lambda m}.$$

Formula (39) implies

$$\int_0^\infty dx \int_0^{x/(r-1)} p(\tau,x)d\tau dx = \lambda \int_0^\infty dx \int_0^{x/(r-1)} q(x - r\tau + \tau)[1 - B(r\tau)]d\tau$$

$$= \frac{\lambda}{r-1} \int_0^\infty dx \int_0^x q(u-s)\left[1 - B\left(\frac{rs}{r-1}\right)\right] ds$$

$$= \frac{\lambda m}{r-1} \int_0^\infty dx \int_0^x q(u-s) g(s) ds - \frac{\lambda m}{r-1} \bar{q}(0) \bar{g}(0)$$

$$= \frac{\lambda^2 m^2 Q(0)}{(r-1)(1-\lambda m)} .$$

Finally,

$$\int_0^\infty n(\tau) d\tau = \lambda Q(0) \int_0^\infty \left[1 - B(r\tau)\right] d\tau = \frac{\lambda Q(0) m}{r-1} ,$$

and normalization gives

$$Q(0)^{-1} = 1 + \frac{\lambda m}{1 - \lambda m} + \frac{(\lambda m)^2}{(r-1)(1 - \lambda m)} + \frac{\lambda m}{r-1} ,$$

or

(43)
$$Q(0) = \frac{1 - \lambda m}{1 + \lambda m/(r-1)} .$$

From the form of $Q(0)$ we take as the condition for equilibrium

$$\lambda m < 1.$$

By taking the limit $r \to \infty$, $q(x)$ and $Q(0)$ approach the classical results for the Takàcs $M/G/1$ model [Tak], as they should.

Example: Let item sizes be distributed exponentially with mean $1/\mu$. Then F is also exponential with failure rate $\nu = 1/m = \mu r/(r-1)$. From (41) we have

$$\bar{q}(s) = \lambda Q(0) \frac{1 - \nu/(\nu+s)}{s - \lambda[1 - \nu/(\nu+s)]} = \lambda Q(0) \frac{s}{s(\nu+s) - \lambda s}$$
$$= \frac{\lambda Q(0)}{s+\nu-\lambda} .$$

Thus,

$$q(x) = \lambda Q(0) e^{-(\nu-\lambda)x} ,$$

where $Q(0)$ is calculated from (43) with $m = (r-1)/r\mu$.

Flow-through and finite capacity, reflection protocol - The solutions for p and n have the same form as before (see (38) and (39)), and by substituting into (10) and (16) we find

$$(44) \qquad 0 = \frac{dq}{dx} - \lambda q + \lambda \int_0^x f(x-u)q(u)du + \lambda f(x)Q(0),$$

and the boundary condition

$$(45) \qquad q(c) = \lambda \int_0^c q(c-x)[1 - F(s)]ds + \lambda Q(0)[1 - F(c)],$$

with f and F as defined in the previous subsection. To this must be added the boundary condition (see (13)), $q(0) = \lambda Q(0)$, and the normalization that determines $Q(0)$:

$$1 = Q(0) + \int_0^c q(x)dx + \int_0^{c} \int_0^{x/(r-1)} p(\tau,x)d\tau dx + \int_0^{c/(r-1)} n(\tau)d\tau.$$

Example 1: Assume exponential item sizes with mean $1/\mu$. Then F is also exponential with failure rate

$$\nu = \frac{\mu r}{r-1}.$$

Equations (44) and (45) give

$$0 = \frac{dq}{dx} - \lambda q + \lambda \nu \int_0^x e^{-\nu(x-u)}q(u)du + \nu Q(0)e^{-\nu x}$$

and

$$q(c) = \lambda e^{-\nu c} \int_0^c e^{\nu u}q(u)du + \lambda Q(0)e^{-\nu c}.$$

By letting

$$u(x) = Q(0) + \int_0^x e^{\nu s}q(s)ds$$

we obtain the boundary value problem

$$u'' - (\nu+\lambda)u' + \nu\lambda u = 0, \ 0 < x < c,$$
$$u(0) = Q(0), \ u'(c) = \lambda u(c).$$

If $\lambda \neq \nu$, then two linearly independent solutions are $e^{\lambda x} \pm e^{\nu x}$ and routine calculations lead to the solution

$$u(x) = Q(0)e^{\lambda x},$$
$$q(x) = \lambda Q(0)e^{(\lambda-\nu)x},$$

$$Q(0) = \cfrac{1}{1 + \cfrac{r}{r-1}\lambda\cfrac{e^{(\lambda-\nu)c}-1}{\lambda-\nu}}.$$

As $c \to \infty$, $Q(0) \to 0$ if $\lambda > \nu$ and $Q(0) \to (1 - \lambda m)/[1 + \lambda m/(r-1)]$ if $\lambda < \nu$, as expected.

Example 2: Let item sizes be uniform on $(0,T)$, $T > rc/(r-1)$. Then $f(s) = r/T(r-1)$ on the interval $(0,c)$, and from (44) and (45) we have

$$0 = \frac{dq}{dx} - \lambda q + \frac{\lambda r}{T(r-1)}\int_0^x q(u)du + \frac{\lambda r}{T(r-1)}Q(0)$$

with the boundary condition,

$$q(c) = \lambda\int_0^c q(c-s)[1 - rs/T(r-1)]ds + \lambda Q(0)[1 - rc/T(r-1)].$$

The other boundary condition is $q(0) = \lambda Q(0)$ as before. With $\alpha = r/T(r-1)$ and

$$Q(x) = Q(0) + \int_0^x q(s)ds$$

we obtain the boundary value problem

(46) $$Q'' - \lambda Q' + \lambda\alpha Q = 0,$$
$$Q'(0) = \lambda Q(0),$$
$$Q'(c) = \lambda(1 - \alpha c)Q(0) + \lambda\int_0^c [1 - \alpha(c-s)]d[Q(x) - Q(0)].$$

The first integration of (46) reveals that the last condition is redundant, as is to be expected; we find easily for $\lambda > 4\alpha$

$$Q(x) = \frac{Q(x)}{\gamma_+ - \gamma_-}\left[(\lambda - \gamma_-)e^{\gamma_+ x} - (\lambda - \gamma_+)e^{\gamma_- x}\right],$$

where

$$\gamma_\pm = \frac{\lambda}{2} \pm \sqrt{\frac{1}{4}\lambda^2 - \lambda\alpha}\,.$$

It is easily verified that this solution is positive and increasing. For $\lambda < 4\alpha$ the roots are

$$\gamma_\pm = \frac{\lambda}{2} \pm i\theta$$

with $2\theta = \sqrt{4\lambda\alpha - \lambda^2}$. We find

$$q(x) = Q'(x) = Q(0)\{\lambda\cos\theta x + \left[\frac{\lambda^2}{4\theta} - \theta\right]\sin\theta x\}e^{\frac{1}{2}\lambda x}.$$

We now prove that $q(x) > 0$ when $\alpha c < 1$. Since $q(0) > 0$, it is sufficient to prove that $q(x) \neq 0$ for $x > 0$. We consider three cases: (i) $2\alpha < \lambda < 4\alpha$, (ii) $\alpha \leqslant \lambda \leqslant 2\alpha$, and (iii) $\lambda < \alpha$.

(i) If for some $x \in (0,c]$ $q(x) = 0$ then $\tan\theta x = 2\theta/(2\alpha - \lambda) < 0$. On the other hand, $0 < \theta x \leqslant \alpha x \leqslant \alpha c < 1$ and we have a contradiction.

(ii)   If   for   some $x \in (0,c]$ $q(x) = 0$,   then $\tan\theta x = 2\theta/(2\alpha - \lambda)$ $= 2\theta c/(2\alpha c - \lambda c) \geqslant 2\theta c$. On the other hand, $\tan\theta x \leqslant \theta c\tan 1$ for $0 \leqslant x \leqslant c$, since $\theta x \leqslant \alpha c < 1$, and again we have a contradiction.

(iii) We take $\lambda = \alpha\beta$ where $0 < \beta < 1$. Then $\theta = \alpha(4\beta - \beta^2)^{1/2}/2$. Further, if for some $x \in (0,c]$, $q(x) = 0$, then

$$\tan\left[\frac{\alpha(4\beta - \beta^2)^{1/2}}{2}x\right] = \frac{(4\beta - \beta^2)^{1/2}}{2 - \beta}\,.$$

Now we prove that in fact

$$\tan\left[\frac{\alpha(4\beta - \beta^2)^{1/2}}{2}x\right] < \frac{(4\beta - \beta^2)^{1/2}}{2 - \beta}\,, \quad 0 < x \leqslant c\,.$$

We put $\delta^2 = 4\beta - \beta^2$. Then $\beta = 2 - \sqrt{4 - \delta^2}$, and since $\alpha x < 1$ for $x \in (0,c]$, it is sufficient to prove that

$$\tan\frac{\delta}{2} < \frac{\delta}{\sqrt{4-\delta^2}} \ , \ \ 0 < \delta < 2.$$

We have

$$\tan^2(\delta/2) < \delta^2/(4-\delta^2) =- 1 + 1/(1-\delta^2/4),$$

or $1/\cos^2(\delta/2) < 1/(1-\delta^2/4)$, or $\sin\delta/2 < \delta/2$ for $0 < \delta < \pi$. This completes the proof.

Flow-through, non-overflowing protocol - Equation (22) with initial condition (24) is solved by characteristics and variation of parameters to give

$$p(x,z) = p(x + \frac{r-1}{r}z, 0) - \frac{\lambda}{r}\int_0^z q\left[x + \frac{r-1}{r}(z-s)\right]b(s)ds.$$

Setting $x = 0$ and $[(r-1)/r]z = x$ produces

$$p(0,\frac{rx}{r-1}) = p(x,0) - \frac{\lambda}{r}\int_0^{\frac{rx}{r-1}} q(x - \frac{r-1}{r}s)b(s)ds.$$

Substituting $p(x,0)$ from the last equation into (21), we have

$$0 = \frac{dq}{dx} - \lambda B\left[\frac{r(c-x)}{r-1}\right]q + \frac{\lambda r}{r-1}\int_0^x q(x-u)b\left[\frac{ru}{r-1}\right]du + \lambda Q(0)b\left[\frac{rx}{r-1}\right]\frac{r}{r-1},$$

and with the rescaled density

$$f(s) = \frac{r}{r-1}b\left[\frac{rs}{r-1}\right]$$

and $Q(x) = Q(0) + \int_0^x q(s)ds,$

$$Q'' - \lambda F(x)Q' + \lambda\int_0^x f(x-u)d(Q(u) - Q(0)) + \lambda Q(0)f(x).$$

The boundary condition $Q'(0) = \lambda Q(0)$ is implied by (23).

Example. Let item sizes be uniform over $(0,T)$, $T > \dfrac{rc}{r-1}$. Then with

$f(x) = \dfrac{r}{T(r-1)} \equiv \alpha$ on $(0,c)$, and $F(x) = \alpha x$, we have the boundary value problem

$$Q'' - \lambda\alpha(c-x)Q' + \lambda\alpha Q = 0 \ , \ Q'(0) = \lambda Q(0).$$

As observed by Beneš [Ben1] the linear term suggests a solution in terms of parabolic cylinder functions, since the change

$$Q(x) = v(x)e^{\frac{1}{2}\alpha\lambda(cx - \frac{x^2}{2})} \ , \ z = \sqrt{\lambda\alpha}(x-c)$$

gives the equation

$$v'' - \left[\frac{z^2}{4} - \frac{1}{2}\right]v = 0$$

with the linearly independent solutions

$$U\left(-\frac{1}{2},z\right) = e^{-\frac{z^2}{4}},$$

(47)
$$V\left(-\frac{1}{2},z\right) = \sqrt{\frac{2}{\pi}}e^{-\frac{z^2}{4}}\int_0^z e^{-t^2/2}dt$$

as in [Hand].

Loading followed by transfer - Equation (29) with the initial condition (34) is solved by the method of characteristics and yields

$$p(x,y,z) = p(x+y/r, 0, z+y) - \frac{\lambda}{r}q(x+y/r)b(z+y).$$

Similarly, (30) with the initial condition (35) yields, using a variation-of-parameters term,

$$n(y,z) = n(0,y+z) + \frac{1}{r}\int_0^y p(0,s,y+z-s)ds$$

$$= \frac{\lambda}{r}Q(0)b(y+z) + \frac{\lambda}{r^2}\int_0^y q(s/r)b(y+z)ds.$$

Substitution into (31) then leads to an equation with the single unknown q,

$$0 = \frac{dq}{dx} - \lambda q + \lambda \int_0^x q\left[x/r + \frac{r-1}{r}u\right]b(x-u)\,du$$

$$+ \lambda b(x)\left[Q(0) + \int_0^{x/r} q(s)\,ds\right].$$

As the loading rate $r \to \infty$ we obtain an instantaneous loading model with Poisson input; in this limit the equation tends to the classical Takàcs equation [Tak] for the density.

The appearance of the retarded argument, x/r, makes the equation difficult to solve, except in special cases. A tractable special case of some interest arises when $r = 1$, for then there is no retarded argument, and the equation becomes

$$0 = \frac{dq}{dx} - \lambda q + \lambda B(x)q + \lambda b(x)\left[q + \int_0^x q(s)\,ds\right].$$

Here we can set

$$Q(x) = Q(x) + \int_0^x q(s)\,ds$$

and find that Q satisfies

$$0 = Q'' - \lambda\left[1 - B(x)\right]Q' + \lambda b(x)Q$$

with the boundary condition $Q'(0) = \lambda Q(0)$ (see (36)). Using $b(x) = B'(x)$ a first integration yields

$$Q' - \lambda\left[1 - B(x)\right]Q = \text{constant} = Q'(0) - \lambda Q(0) = 0,$$

whereupon a second integration results in

$$Q(x) = Q(0)\exp\left\{\lambda\int_0^x [1 - B(s)]\,ds\right\},$$

from which q, n, and p are determined readily. It remains to find $Q(0)$ from the normalization condition

$$1 = Q(0) + \int_0^\infty q(x)dx + \int_0^\infty \int_0^\infty n(y,z)dydz + \int_0^\infty \int_0^\infty \int_0^\infty p(x,y,z)dxdydz.$$

Since

$$\int_0^\infty q(s)ds = Q(\infty) - Q(0) = Q(0)\left[e^{\lambda b} - 1\right],$$

$$\int_0^\infty \int_0^\infty n(y,z)dydz = \lambda \int_0^\infty [1-B(y)]Q(y)dy,$$

$$\int_0^\infty \int_0^\infty \int_0^\infty p(x,y,z)dxdydz = Q(0)\lambda be^{\lambda b} - \lambda \int_0^\infty [1-B(y)]Q(y)dy,$$

we have

$$Q(0) = \frac{e^{-\lambda b}}{1+\lambda b},$$

where $b = \int_0^\infty xb(x)dx$ is the mean item size, so that $\rho = \lambda b$ may be interpreted as the traffic intensity.

We see that the emptiness probability $Q(0)$ is a decreasing function of ρ, but that for equilibrium to exist there is no constraint on ρ. There exists a stationary distribution for all ρ, because the loading and output rates are the same and because items arriving during loading periods are rejected; a large item size incurs a proportionally large loading time during which no new arrivals occur. We note that the emptiness probability $Q(0)$ has an exponential dependence on ρ, while the corresponding dependence in the M/G/1 queue is linear.

Loading with transfer, finite capacity - We have the following modified stationary equations

$$0 = \frac{\partial p}{\partial x} + r\frac{\partial p}{\partial z} - r\frac{\partial p}{\partial y}, \quad 0<z+y\leqslant k(x+y/r),$$

$$0 = -r\frac{\partial n}{\partial y} + r\frac{\partial n}{\partial z} + p(0,y,z), \quad 0<z+y\leqslant c,$$

$$0 = \frac{dq}{dx} - \lambda B(k(x))q + \int_{0\leqslant y\leqslant k(x+y/r)} p(x-y,y,0)dy + rn(x,0), \ 0<x\leqslant c,$$

with the boundary conditions

$$rp(x,0,z) = \lambda q(x)b(z) , \quad 0 \leqslant z \leqslant k(0),$$
$$rn(0,z) = \lambda Q(0)b(z) , \quad 0 \leqslant z \leqslant k(x),$$
$$q(0) = \lambda Q(0).$$

Solving as before leads to

$$p(x,y,z) = \frac{\lambda}{r}q(x+y/r)b(z+y) , \quad z+y \leqslant k(x+y/r)$$

$$n(y,z) = \frac{\lambda}{r}b(y+z)\left[Q(0) + \int_0^{y/r}q(s)ds\right] , \quad 0 \leqslant y+z \leqslant c,$$

which in turn yields the single equation

$$0 = \frac{dq}{dx} - \lambda B(k(x))q + \lambda \int_{0 \leqslant y \leqslant k(x+y/r)} q(x-y+y/r)b(y)dy$$

$$+ \lambda b(x)\left\{Q(0) + \int_0^{y/r}q(s)ds\right\}, \quad 0 < x \leqslant c,$$

together with the boundary condition $q(0) = \lambda Q(0)$ and the normalization

$$1 = Q(0) + \int_0^c q(x)dx + \iint_{0 \leqslant y+z \leqslant c} n(y,z)dydz + \iiint_{0 \leqslant y+z \leqslant k(x+y/r)} p(x,y,z)dxdydz.$$

For $r = 1$, the distribution $Q(x) = Q(0) + \int_0^x q(s)ds$ satisfies the boundary value problem

$$0 = Q'' - \lambda\left[B(k(x)) - \int_{0 \leqslant y \leqslant k(x+y)} dB(y)\right]Q' + \lambda b(x)Q,$$

$$Q'(0) = \lambda Q(0).$$

Example: Arriving items are uniformly distributed on $[0,T]$, with $T > c$, and the lazy protocol $k(x) = c-x$ is used. Then $B(x) = x/T$ and $b(x) = 1/T$ over the range of interest, and

$$B(k(x)) - \int_{0 \leqslant y \leqslant k(x+y)} dB(y) = \frac{c-x}{2T},$$

$$Q'' - \frac{\lambda}{2T}(c-x)Q' + \frac{\lambda}{T}Q' = 0 \ , \ \ 0<x\leqslant c, \ Q'(0) = \lambda Q(0).$$

As shown by Beneš [Ben1] the transformation

$$Q(x) = v(z)\exp\left\{\frac{\lambda}{4T}(cx - \frac{1}{2}x^2)\right\}, \ z = \sqrt{\lambda/2T}\,(x-c)$$

produces

$$v'' - \left[\frac{z^2}{4} - \frac{3}{2}\right]v = 0,$$

which has the parabolic cylinder functions

$$U\left(-\frac{3}{2}, z\right) = ze^{-z^2/4}$$

and

$$V\left(-\frac{3}{2}, z\right) = \sqrt{\frac{2}{\pi}}\left[ze^{-z^2/4}\int_0^z e^{u^2/2}du - e^{z^2/4}\right]$$

as linearly independent solutions (see [Hand] and note that $V\left(-\frac{1}{2}, z\right)$ is given by (47)).

4. Infinite Buffer with Multiple Sources

In this section we consider a model analyzed by Anick, Mitra and Sondhi [AniMS]. It consists of an infinite capacity buffer that receives items (messages) from a finite number, N, of information sources, which independently and asynchronously alternate between the transmitting and idle states. The transmission periods as well as the idle periods are exponentially distributed for each source. While the parameters of these two distributions are different in general, they are common to all sources; also the sources are mutually independent. The average idle period is denoted $1/\lambda$. Without loss of generality, we adopt the following normalization. The unit of time is selected to be the average transmission period and the unit of information is chosen to correspond to an item of average length. In these units the sources transmit at a uniform rate of 1 unit of information per unit of time. Thus, when r sources are active simultaneously, the instantaneous receiving rate at the buffer is r. The buffer accumulates information when the receiving rate exceeds the maximum transmission rate, r_0, of the output channel. (Note that by our normalization, r_0 is also the ratio of the output channel capacity to an active source's transmission rate.) We assume that $r_0 < N$, since otherwise the buffer is always empty.

As long as the buffer is not empty, the instantaneous rate of change of the buffer content is $r-r_0$, i.e. we have loading with flow-through. Once the buffer is empty, it remains so as long as $r \leqslant r_0$. We assume that the buffer is infinite and that the following stability condition is satisfied:

$$(48) \qquad \qquad \frac{N\lambda}{r_0(1+\lambda)} < 1.$$

In [AniMS] the reader will find a brief discussion of the application of the above model to the study of switches in computer networks. The input devices may range from slow work-station terminals to high data rate computer systems.

Equilibrium equations - If at time t the number of active sources is i, the two elementary events, start-up of a new source and turn-off of an active source in the interval $[t, t+\Delta t]$, have the respective probabilities $i\Delta t + o(\Delta t)$ and $(N-i)\Delta t + o(\Delta t)$. Here, of course, we have used the exponential assumptions for the transmission and idle periods. In $[t, t+\Delta t]$ multiple events have probabilities $o(\Delta t)$, and the probability of no change is $1 - \{(N-i)\lambda + i\}\Delta t + o(\Delta t)$.

Let $Q_i(x,t)$, $0 \leqslant i \leqslant N$, $t \geqslant 0$, $x \geqslant 0$, be the probability that at time t, i sources are active and the buffer content is no greater than x. From the above observations

$$Q_i(x,t+\Delta t) = \{N - (i-1)\}\lambda\Delta t Q_{i-1}(x,t) + (i+1)\Delta t Q_{i+1}(x,t)$$
$$+ [1 - \{(N-i)\lambda+i\}\Delta t]Q_i(x-(i-r_0)\Delta t,t) + o(\Delta t), \; 0 \leqslant i \leqslant N.$$

Passing to the limit $\Delta t \to 0$, we obtain

$$(49) \quad \frac{\partial Q_i(x,t)}{\partial t} + (i-r_0)\frac{\partial Q_i(x,t)}{\partial x} =$$
$$(N-i+1)\lambda Q_{i-1}(x,t) - \{(N-i)\lambda+i\}Q_i(x,t) + (i+1)Q_{i+1}(x,t), \; 0 \leqslant i \leqslant N.$$

Our primary objective is the stationary probability distribution $\{Q_i(x)\}$,

$$Q_i(x) = \lim_{t \to \infty} Q_i(x,t), \, 0 \leqslant i \leqslant N.$$

Therefore, taking the limit $t \to \infty$ in (49) we eliminate the derivative with respect to t and get for $i = 0,1, \ldots, N$,

(50) $\qquad (i-r_o) \dfrac{dQ_i(x)}{dx} =$

$$(N-i+1)\lambda Q_{i-1}(x) - \{(N-i)\lambda+i\}Q_i(x) + (i+1)Q_{i+1}(x),$$

where it is understood that $Q_i(x) \equiv 0$ if $i < 0$ or $i > N$. In matrix notation,[†] with the column vector $\vec{Q}(x) = [Q_o(x), \ldots, Q_N(x)]'$, we have

(51) $\qquad\qquad \mathbf{D}\dfrac{d}{dx}\vec{Q}(x) = \mathbf{M}\vec{Q}(x) \, , \, x \geqslant 0,$

where $\mathbf{D}$ is a diagonal matrix with elements $d_{ii} = i-r_o, \, 0 \leqslant i \leqslant N$, and where

$$\mathbf{M} = \begin{bmatrix} N\lambda & 1 & 0 & 0 & & & \\ -N\lambda & -\{(N-1)\lambda+1\} & 2 & 0\ldots & & & \\ 0 & (N-1)\lambda & -\{(N-2)\lambda+2\} & 3 & & & \\ & \vdots & & & & & \\ & & & & & & 0 \\ & & & & \ldots\, 2\lambda & -\{\lambda+(N-1)\} & N \\ & & & & & \lambda & -N \end{bmatrix}$$

When r_o is an integer, one of the differential equations in (50) degenerates to an algebraic equation that may be used to eliminate one of the unknown components of $\vec{Q}(\cdot)$. Thus, in the analysis to follow, we assume that r_o is not an integer.

To specify the initial conditions for the differential equations, we first note that if the number of sources at any time exceeds r_o, then the buffer content increases

---

† Boldface denotes matrices, arrows denote vectors and a prime denotes transposition.

and the buffer cannot stay empty. It follows that

$$(52) \qquad\qquad Q_i(0) = 0, \ \lfloor r_o \rfloor + 1 \leqslant i \leqslant N,$$

where $\lfloor r_o \rfloor$ denotes the integer part of r_o.

By supplementing (52) with the tri-diagonal structure of the matrix $\mathbf{D}^{-1} \mathbf{M}$ we can further characterize the behavior of $\vec{Q}(x)$ when x is small. Observe that an application of $\mathbf{D}^{-1} \mathbf{M}$ to $\vec{Q}(0)$ decreases by 1 the number of trailing zero elements, and that each additional application has the same effect until $(\mathbf{D}^{-1} \mathbf{M})^{N-\lfloor r_o \rfloor - 1} \vec{Q}(0)$ has only its last component equal to zero and $(\mathbf{D}^{-1} \mathbf{M})^{N-\lfloor r_o \rfloor}$ has none. Thus

$$(53) \qquad\qquad \{(\mathbf{D}^{-1} \mathbf{M})^j \vec{Q}(0)\}_i = 0, \ \lfloor r_o \rfloor + 1 + j \leqslant i .$$

Now recall that because of the differential equations governing $\vec{Q}(\cdot)$ in (51),

$$\vec{Q}^{(j)}(0) = (\mathbf{D}^{-1} \mathbf{M}) \vec{Q}^{(j-1)}(0) = (\mathbf{D}^{-1} \mathbf{M})^j \vec{Q}(0) .$$

Thus from (53) we find

$$Q_i^{(j)}(0) = 0, \quad \lfloor r_o \rfloor + 1 + j \leqslant i ,$$

and in particular the following useful relation,

$$(54) \qquad\qquad Q_N^{(j)}(0) = 0, \quad j = 0, 1, \ldots, N - \lfloor r_o \rfloor - 1 .$$

Thus, not only does the event "all sources are active and the buffer is empty" have probability zero, a fact already known from (52), but also the growth of the probability is slow when the buffer content is small.

Eigenvalues and eigenvectors - If the matrix $\mathbf{D}^{-1} \mathbf{M}$ has $N + 1$ distinct eigenvalues $z_0, z_1, \ldots, z_N$, then a general solution to the differential equation in (51) can be written as

$$\vec{Q}(x) = \sum_{i=0}^{N} C_i e^{z_i x} \vec{\phi}_i ,$$

where $\vec{\phi}_i = (\phi_{i0}, \phi_{i1}, \ldots, \phi_{iN})$, $0 \leqslant i \leqslant N$, are the associated right eigenvectors and the C_i's are arbitrary constants. In this subsection we develop the procedure for obtaining the eigenvalues and the eigenvectors.

Let z be some eigenvalue of $\mathbf{D}^{-1}\mathbf{M}$ and let $\vec{\phi}$ be the associated right eigenvector. That is,

$$(55) \qquad\qquad z\mathbf{D}\vec{\phi} = \mathbf{M}\vec{\phi}.$$

Equation (55) can also be rendered as

$$(56) \quad z(i-r_o)\phi_i = \lambda(N+1-i)\phi_{i-1} - \{(N-i)\lambda + i\}\phi_i + (i+1)\phi_{i+1}, \quad 0 \leqslant i \leqslant N.$$

Define the generating function of $\vec{\phi}$,

$$(57) \qquad\qquad \Phi(x) = \sum_{i=0}^{N} \phi_i x^i.$$

After multiplying (56) by x^i and summing over i, we expect to obtain an equation in $\Phi(x)$ and $\Phi'(x)$, since $\sum i x^i \phi_i = x\Phi'(x)$. In fact, we get

$$(58) \qquad\qquad \frac{\Phi'(x)}{\Phi(x)} = \frac{zr_o - N\lambda + N\lambda x}{\lambda x^2 + (z+1-\lambda)x - 1}.$$

We solve this differential equation in terms of the distinct real roots, x_1 and x_2, $x_1 > 0 > x_2$, of the quadratic in the denominator of the right-hand side, i.e.

$$(59a) \qquad x_1 = \{-(z+1-\lambda) + \sqrt{(z+1-\lambda)^2+4\lambda}\}/2\lambda,$$
$$(59b) \qquad x_2 = \{-(z+1-\lambda) - \sqrt{(z+1-\lambda)^2+4\lambda}\}/2\lambda.$$

In terms of x_1 and x_2 (58) may now be expressed as

$$(60) \qquad\qquad \frac{\Phi'(x)}{\Phi(x)} = \frac{b_1}{x-x_1} + \frac{b_2}{x-x_2},$$

where a calculation of the residues yields

$$(61a) \qquad\qquad b_2 = N - b_1,$$
$$(61b) \qquad\qquad b_1 = \frac{zr_o - N\lambda + N\lambda x_1}{\lambda(x_1-x_2)}.$$

The solution to (60) is

$$(62) \qquad\qquad \Phi(x) = (x-x_1)^{b_1}(x-x_2)^{N-b_1},$$

where, as in the rest of this section, we have assumed $\phi_N = 1$.

We now make an observation about (62) which is crucial to the present derivation. Note that by its definition in (57), $\Phi(x)$ is a polynomial in x of degree N. Since x_1 and x_2 are distinct, this is possible if and only if b_1 defined in (61b) is an integer in $[0, N]$. Denoting this integer by k we get

(63) $$\Phi(x) = (x - x_1)^k (x - x_2)^{N-k}, \ k = 0, 1, \ldots, N.$$

In (61b) write k for b_1 and use (59) to substitute expressions for x_1 and $x_1 - x_2$. After rearranging and squaring we obtain the following family of quadratics in the unknown eigenvalue z,

$$A(k)z^2 + B(k)z + C(k) = 0, \ k = 0, 1, \ldots, N,$$

where the coefficients are defined by

(64)
$$
\begin{aligned}
A(k) &= (N/2 - k)^2 - (N/2 - r_o)^2, \\
B(k) &= 2(1 - \lambda)(N/2 - k)^2 - N(1+\lambda)(N/2 - r_o), \\
C(k) &= -(1+\lambda)^2 \{(N/2)^2 - (N/2 - k)^2\}.
\end{aligned}
$$

To summarize, we have shown that all the roots of the above family of $N + 1$ quadratics are eigenvalues, as defined in (55). We denote by $z_1^{(k)}$ and $z_2^{(k)}$ the two roots of the kth quadratic. The theorem below, which is taken from [AniMS] is a collection of various properties of the roots.

Theorem [AniMS]

(i) The quadratics for k and k' are identical when $N/2 - k = k' - N/2$.

(ii) For each $k < N/2$ the corresponding quadratic has two real and simple roots. When N is even and $k = N/2$, the corresponding quadratic has a repeated real root.

(iii) If $k' < k \leqslant N/2$ then

$$A(k')z^2 + B(k')z + C(k') > A(k)z^2 + B(k)z + C(k), \text{ for all } z .$$

(iv) The roots of the quadratic corresponding to any k are distinct from those of the quadratic corresponding to any k', provided $k' < k \leqslant N/2$.

(v) Ignoring the multiplicities, there are $N - \lfloor r_o \rfloor$ negative roots, 1 root at 0 and $\lfloor r_o \rfloor$ positive roots.

(vi) The set of eigenvalues coincides exactly with the set of roots of the quadratics.

(vii) The largest negative eigenvalue is $-(1+\lambda-N\lambda/r_o)/(1-r_o/N)$.

The results of the theorem are derived from an analysis of the explicit forms for the coefficients $A(k)$, $B(k)$ and $C(k)$ given in (64). Details can be found in [AniMS].

We adopt the following convention for the eigenvalues.

$$z_{N-\lfloor r_o \rfloor - 1} < ... < z_1 < z_0 < z_N = 0 < z_{N-1} < ... < z_{N-\lfloor r_o \rfloor}.$$

With this convention, z_k and z_{N-k} are roots of the k-th quadratic, i.e.

$$\left\{ z_k, z_{N-k} \right\} = \left\{ z_1^{(k)}, z_2^{(k)} \right\}.$$

The stable or negative eigenvalues are needed later. For this purpose, the roots are given explicitly below. First, let N be odd and adopt the notational convention, $z_1^{(k)} < z_2^{(k)}$. The stable subset is given below in braces.

If $r_o < N/2$, then

(65a)

$$\left\{ \begin{array}{l} z_1^{(k)} = \left[-B(k) - \sqrt{B^2(k) - 4A(k)C(k)} \right] / 2A(k), \quad 0 \leqslant k \leqslant \lfloor r_o \rfloor, \\ z_{1,2}^{(k)} = \left[-B(k) \mp \sqrt{B^2(k) - 4A(k)C(k)} \right] / 2A(k), \quad \lfloor r_o \rfloor + 1 \leqslant k < N/2, \end{array} \right\}$$

and if $r_o > N/2$, then

(65b)

$$\left\{ z_1^{(k)} = \left[-B(k) - \sqrt{B^2(k) - 4A(k)C(k)} \right] / 2A(k), \quad 0 \leqslant k \leqslant N - \lfloor r_o \rfloor - 1 \right\}.$$

When N is even, the only change is that the expressions in (65a) are augmented by $-N(1+\lambda)/(n-2r_o)$, which is one of the repeated roots of the quadratic for $k = N/2$.

We describe next the procedure for obtaining the eigenvectors from the eigenvalues. Given an eigenvalue z, we compute in order, using (59) and (61), the quantities x_1, x_2, and k. These are used in (63) to produce the coefficients of the given polynomial and thus the eigenvector coefficients. Therefore, for the ith component of the eigenvector we have

$$(66) \qquad \phi_i = (-1)^{N-i} \sum_{j=0}^{k} \binom{k}{j} \binom{N-k}{i-j} x_1^{k-j} x_2^{N-k-i+j}, \quad 0 \leqslant i \leqslant N.$$

A general, bounded solution to the differential equations in (51) can now be written as

$$(67) \qquad \vec{Q}(x) = C_N \vec{\phi}_N + \sum_{i=0}^{N-\lfloor r_0 \rfloor - 1} C_i e^{z_i x} \vec{\phi}_i .$$

Since $Q_i(\infty)$ is the probability that i out of N sources are active simultaneously, we have

$$(68) \qquad Q_i(\infty) = \frac{1}{(1+\lambda)^N} \binom{N}{i} \lambda^i, \quad 0 \leqslant i \leqslant N.$$

Hence, because of our assumption $\phi_{NN} = 1$, we obtain

$$\vec{\phi}_N = \frac{1}{\lambda^N} \left[1, N\lambda, \ldots, \binom{N}{i}\lambda^i, \ldots, \lambda^N \right],$$

and $C_N = \lambda^N/(1+\lambda)^N$.

Next, we evaluate the coefficients, C_i, in the solution expressed by (67). By our convention $\phi_{iN} = 1$, we find from (67)

$$Q_N(x) = \left(\frac{\lambda}{1+\lambda} \right)^N + \sum_{i=0}^{N-\lfloor r_0 \rfloor - 1} C_i e^{z_i x}, \quad x \geqslant 0.$$

The above, together with (54) imply the following set of equations.

$$(69) \qquad \sum_{j=0}^{N-\lfloor r_0 \rfloor - 1} z_j^i C_j = -\left(\frac{\lambda}{1+\lambda} \right)^N \delta_{0i}, \quad 0 \leqslant i \leqslant N - \lfloor r_0 \rfloor - 1,$$

where $\delta_{00} = 1$ and $\delta_{0i} = 0$ for $i \neq 0$. Equation (69) in matrix form is

$$(70) \qquad V\vec{C} = -\left(\frac{\lambda}{1+\lambda} \right)^N \vec{e}$$

where $V_{ij} = z_j^i$, $\vec{C} = (C_0, C_1, \ldots, C_{N-\lfloor r_o \rfloor - 1})'$ and $\vec{e} = (1, 0, \ldots, 0)'$.

Recognizing V as a Vandermonde matrix, we may exploit well-known results on such matrices to solve (70) explicitly. Note that V is nonsingular because the eigenvalues $\{z_i\}$ are distinct [Bell], as previously established in the theorem. Therefore,

$$|V| = \prod_{0 \leqslant i \leqslant j \leqslant N - \lfloor r_o \rfloor - 1} (z_i - z_j) .$$

This formula, applied to the minors, which also are related to Vandermonde matrices, gives

(71) $$C_j = -\left(\frac{\lambda}{1+\lambda}\right)^N \prod_{\substack{i=0 \\ i \neq j}}^{N-\lfloor r_o \rfloor - 1} \frac{z_i}{z_i - z_j} , \quad 0 \leqslant j \leqslant N - \lfloor r_o \rfloor - 1.$$

To summarize, the above procedure for obtaining $\vec{Q}(x)$ is based on using the expression (67), where $C_N \vec{\phi}_N = \vec{Q}(\infty)$ is given by (68). Only the stable eigenvalues appear in the above form and they are given explicitly in (65); $\{\phi_i\}$ is obtained from the generating function in (62) and the coefficients C_i appear in (71).

Probability of overflow, moments and asymptotics - We define[†]
$$G(x) = Pr\{\text{buffer content} > x\} = 1 - \vec{1}'\vec{Q}(x), \quad x \geqslant 0$$
as the probability of overflow beyond x. Using the results of the previous subsection we have (recall that $\sum_{i=0}^{N} Q_i(\infty) = 1$)

(72) $$G(x) = -\sum_{i=0}^{N-\lfloor r_o \rfloor - 1} C_i (\vec{1}'\vec{\phi}_i) e^{z_i x} .$$

Given the elements of the series solution in (67), expressions for the moments of the equilibrium buffer content are easily computed. Thus, for the n-th moment

$$m_n = \int_0^\infty x^n d(\vec{1}'\vec{Q}(x)) = n \int_0^\infty x^{n-1} G(x) dx,$$

we use (72) for $G(x)$ and obtain

(73)
$$m_n = \frac{n!}{(-1)^{n+1}} \sum_{i=0}^{N-\lfloor r_o \rfloor -1} \frac{C_i(\overrightarrow{1'\phi_i})}{z_i^n}.$$

Note that $\overrightarrow{1'\phi} = \Phi(1)$, which simplifies the computation in (72) and (73).

Finally, we examine the asymptotic behavior of the probability of overflow beyond x. The asymptotic formulas for large x are useful for two reasons: They often describe system behavior rather well in all but those regions of least importance, where x is small; and the analytic formulas are simple, easily interpreted, and exhibit the essential behavior.

Since the form of the solution in (67) is a sum of exponential terms and $C_N \overrightarrow{\phi_N} = \overrightarrow{Q}(\infty)$, the error in approximating $Q(x)$ by $\overrightarrow{Q}(\infty)$ will be dominated by the term with the largest exponent. Hence[t]

$$\overrightarrow{Q}(x) - \overrightarrow{Q}(\infty) \sim C_0 \overrightarrow{\phi_0} e^{-\theta x},$$

and

$$G(x) \sim -C_0 (\overrightarrow{1'\phi_0}) e^{-\theta x},$$

where $-\theta = z_0$ is the largest negative eigenvalue of $\mathbf{D}^{-1}\mathbf{M}$. Using statement (vii) of the theorem, a calculation of θ yields

(74)
$$\theta = \frac{1+\lambda-N\lambda/r_o}{1-r_o/N} = \frac{(1+\lambda)(1-\rho)}{1-r_o/N}$$

where in the latter expression ρ is the traffic intensity given in the left hand side of inequality (48). The coefficient C_0 is given in (71) and the eigenvector $\overrightarrow{\phi_0}$ is obtained by the general procedure that leads to (66). We find that

$$x_1 = 1 - \frac{N}{r_o}, \quad x_2 = \frac{1}{\lambda} \cdot \frac{1}{N/r_o-1}, \quad k=N,$$

$$\Phi(x) = \left\{ x + \left[\frac{N}{r_o} - 1 \right] \right\}^N.$$

---

† Let $\overrightarrow{1}$ denote a row vector of all 1's.

† By $a(x) \sim b(x)$ we mean that $a(x)/b(x) \to 1$ as $x \to \infty$.

Hence,

$$\phi_{0i} = \binom{N}{i} \left[\frac{N}{r_o} - 1 \right]^{N-i} , \quad 0 \leqslant i \leqslant N,$$

and

$$\vec{1}'\vec{\phi}_0 = \left[\frac{N}{r_o} \right]^N .$$

Collecting terms we conclude that

(75) $$G(x) \sim \rho^N \left\{ \prod_{i=1}^{N-\lfloor r_o \rfloor - 1} \frac{z_i}{z_i + \theta} \right\} e^{-\theta x} .$$

Final remarks. The model analyzed in this section is closely related to a model studied by Kosten [Kos2]. The latter model is in fact simply the limiting case with $N \to \infty$, $\lambda \to 0$ in such a way as to retain a finite traffic intensity ρ. Note also that the sequence of instants when transmission periods begin yields, by assumption, a Poisson process.

Another similar model was analyzed by Gaver and Lewis [GavL]. In their case, the buffer was partitioned into blocks, one for each input source. Also, output was instantaneous for each message, but it could not take place until the end of loading. In their analysis they exploited the normal approximation to the sum of the contents of each source block.

For the finite, multiple source model two principal open questions that remain are the extensions to general distributions of transmission periods and to finite capacity buffers with the various overflow protocols illustrated in earlier sections. For the exponential case and $c = \infty$, a broad collection of numerical results are presented in [AniMS] for the purpose of measuring the interplay between the probability of overflow (grade of service), the number N of sources and the traffic intensity. These results also quantify the approximations inherent in Kosten's $N = \infty$ model and in the asymptotic formula for overflow probabilities (see (74) and (75)).

Chapter 3

Primary Computer Storage

1. Introduction

In the buffer models that were analyzed previously, the storage device was assumed to be depleted at a constant rate when nonempty. This assumption is unsuitable for modeling such systems as computer memories, in which several new features appear. For example, the length of time a program stays in memory, i.e., its execution time, may be correlated with its size, and it may be increased because of other programs that are competing for the processor. Also, the item stored, typically a data set or a program to be executed, retains its identity in the sense that departing items and their sizes are in an obvious one-to-one correspondence with arriving items and their sizes. This "conservation principle" presents a major difficulty in the analysis, for in many Markov models it means that we have to carry along as part of the state variable an indication of the size of each item in memory. These have been termed models of "exact content" by Beneš [Ben1].

As a concession to the greater difficulty of computer memory models, the loading and output processes in buffer models are not brought out explicitly. Thus, from an applications standpoint, our results apply chiefly to those systems in which the effects of loading and unloading are negligible in comparison to those of execution times and the residence times of items in memory. In some instances, however, an approximate accounting for loading and unloading times can be accomplished by including them in the execution or residence times.

The central problem in what follows is to characterize the distribution of the space required by all items present in the system. We begin with a model that takes into account only the space-time correlation. We then consider models of finite memories, a model of storage fragmentation and finally, diffusion approximations for storage processes.

53

2. The M/G/1 Storage Process

In this section we model the storage problem as an $M/G/1$ queue with the first-come first-served (FCFS) discipline. The analysis follows the original work of Sengupta [Sen]. The arrival rate is assumed to be λ, and a service time density $g(t)$ with mean $1/\mu$ is assumed to exist. We assume that the space required by an item is independent of that required by other items, and that it is occupied up to the instant the item completes service. We allow the space, Y, required by an item to depend on its service time, S, in the sense that there is a known joint distribution function

$$g(y, t)dt = Pr\{Y \leq y, t < S \leq t+dt\} .$$

Our goal is the steady state distribution of the total space required by all items present in the queueing system. We give an explicit characterization of the dependence between the space needed by the waiting items and that needed by the item in service.

Consider first the space required by the waiting items. One approach to the problem is based on the treatment of the $M/G/1$ queue by the supplementary variable technique (e.g. see [Coh1] and [Kle]).

We denote by C_W the space occupied by the waiting items and by X the elapsed service time (age) of the item in service. We then define

$$R(y,x)dx = Pr\{\text{Server busy}, C_W \leq y, x \leq X \leq x+dx\} .$$

Let Y_i be the space occupied by the i-th waiting item. The Y_i's are assumed to be i.i.d. random variables with the common distribution function

$$B(y) = Pr\{Y \leq y\} = \int_0^\infty g(y,t)dt .$$

If N_W denotes the number of waiting items, then

$$C_W = \sum_{i=1}^{N_W} Y_i$$

with the Y_i's being independent of N_W.

Next, we introduce

$$m_k(x) = Pr\{\text{Server busy}, N_W = k, x \leqslant X \leqslant x+dx\}$$

and its generating function $M(z,x) = \sum_{k \geqslant 0} m_k(x)z^k$. Thus, denoting Laplace-Stieltjes transforms (LST's) by tildas over the symbols for the corresponding distribution functions, we have

(1) $$\tilde{R}(u,x) = M(\tilde{B}(u),x),$$

and for the transform of the space occupied by waiting items, we have

(2) $$E\left[e^{-uC_W}\right] = \int_0^\infty \tilde{R}(u,x)dx .$$

A routine analysis yields (see [Coh1])

(3) $$M(z,x) = K(z)\left[1-G(x)\right]e^{-\lambda(1-x)},$$

where

$$G(x) = Pr\{S \leqslant x\} = \int_0^x g(u)du ,$$
$$K(z) = \frac{(1-z)\lambda(1-\rho)}{J(z) - z}, \rho = \lambda/\mu,$$

and

$$J(z) = \int_0^\infty e^{-\lambda(1-z)u} g(u)du .$$

Now by (1)–(3) we have routinely $E(C_W) = E(Y) E(N_W)$ and $Var(C_W) = E(N_W) Var(Y) + [E(Y)]^2 Var(N_W)$, where the first relation is also implied by the initial representation of C_W.

Consider next the space, C_S, required by the item in service. We introduce the conditional probability

(4) $$H(y\,|\,x) = Pr\{C_S \leqslant y\,|\,X = x\} = \int_x^\infty g(y,t)dt\,/[1-G(x)]$$

with the transform $\tilde{H}(u\,|\,x)$. The density function describing an item's age is $\mu[1-G(x)]$ and the busy-server probability is ρ; therefore,

$$E(C_S) = -\rho \int_0^\infty \tilde{H}'(0\,|\,x)\mu[1-G(x)]dx = \lambda E(YS) ,$$

(5)
$$E(C_S^2) = \rho \int_0^\infty \tilde{H}''(0|x)\mu[1-G(x)]dx = \lambda E(Y^2 S) .$$

The first result may be justified heuristically by Little's formula. The arrival rate of storage demand is $\lambda E(Y)$; multiplying this by the average time spent by the storage required at the server, viz. $E(YS)/E(Y)$, we obtain $E(C_S)$.

We are now in position to analyze the total space, $C = C_S + C_W$, required by all items. We have

$$Q(y) \equiv Pr\{C \leqslant y\}$$

$$= Pr\{\text{Server idle}\} + \int_0^\infty Pr\{\text{Server busy}, C_S + C_W \leqslant y | X = x\}\mu[1-G(x)]dx .$$

Since N_W (and hence C_W) depends on the item in service only through its age, C_W is independent of C_S, given the age X. Thus, we have the transform

(6)
$$\tilde{Q}(u) = 1 - \rho + \int_0^\infty \tilde{H}(u|x) \tilde{R}(u,x)dx .$$

On applying (1), (3) and (4) a calculation yields

$$\int_0^\infty \tilde{H}(u|x) \tilde{R}(u,x)dx = \int_0^\infty \left[\int_0^\infty e^{-uy} H(dy|x) \right] M(\tilde{B}(u),x)dx$$

$$= \int_0^\infty \left[\int_0^\infty e^{-uy} \left[\int_x^\infty g(dy,t)dt \right] \right] K(\tilde{B}(u)) \exp\{-\lambda(1-\tilde{B}(u))\}dx \, dt$$

$$= \frac{K(\tilde{B}(u))}{\lambda(1-\tilde{B}(u))} \int_0^\infty \int_0^\infty e^{-uy} g(dy,t) \left[1 - \exp\{-\lambda[1-\tilde{B}(u)]t\} \right] dt .$$

Thus,

(7)
$$\int_0^\infty \tilde{H}(u|x) \tilde{R}(u, x)dx = \frac{K(\tilde{B}(u))}{\lambda(1-\tilde{B}(u))} \left[\tilde{B}(u) - V(u, \lambda (1-\tilde{B}(u))) \right] ,$$

where

$$V(u, v) = \int_0^\infty \int_0^\infty e^{-uy} \, e^{-vt} \, g(dy, t)dt .$$

Substituting (7) into (6) we finally have

(8) $$\tilde{Q}(u) = 1 - \rho + \frac{K(\tilde{B}(u))}{\lambda(1 - \tilde{B}(u))} \left[\tilde{B}(u) - V(u, \lambda(1 - \tilde{B}(u))) \right] .$$

A derivative of (8) produces

$$E(C) = -\tilde{Q}'(0) = \lambda E(YS) + E(Y)E(N_W) = E(C_S) + E(C_W) .$$

We conclude by showing that

(9) $$E(C^2) = E(C_W^2) + \lambda E(Y^2 S) + $$
$$2\lambda \, E(N_W) \; E(Y)E(YS) + \lambda^2 E(Y)E(YS^2) ,$$

from which we can determine the correlation coefficient γ between C_S and C_W,

$$\gamma = \frac{\lambda^2 E(Y)E(YS^2)}{2\sqrt{Var(C_S) \; Var(C_W)}} .$$

To derive (9), we use the following relation implied by (6):

(10) $$\tilde{Q}''(u) = \int_0^\infty \tilde{H}''(u \,|\, x) \, \tilde{R}(u, x)dx + \int_0^\infty \tilde{H}(u \,|\, x) \, \tilde{R}''(u, x)dx$$
$$+ 2 \int_0^\infty \tilde{H}'(u \,|\, x) \, \tilde{R}(u, x)dx .$$

The first and second terms of the right hand side can be simplified to

(11) $$\int_0^\infty \tilde{H}''(0 \,|\, x) \, \tilde{R}(0, x)dx = E(C_S^2)$$

(see (5)), and

(12) $$\int_0^\infty \tilde{H}(0 \,|\, x) \, \tilde{R}''(0, x)dx = E(C_W^2) .$$

Further,

(13) $$\tilde{H}'(0|x) - \int_x^\infty \int_0^\infty y g(dy,t)/(1-G(x)) ,$$

and

$$\tilde{R}'(0,x) - [1-G(x)] \exp\{-\lambda(1-\tilde{B}(0))x\} [K'(\tilde{B}(0))\tilde{B}'(0)$$
$$+ K(\tilde{B}(0))\tilde{B}'(0)\lambda x]$$

$$- [1-G(x)][K'(1)E(Y) + \lambda^2 x \, E(Y)] .$$

Next, we use the fact that

$$K(z) \, J(z) - \lambda N(z) ,$$

where $N(z)$ is the generating function of the number of items in the system. We have

$$K'(1) - \lambda N'(1) - \lambda\rho - \lambda E(N_W)$$

and

(14) $$\tilde{R}'(0,x) - [1-G(x)][\lambda E(N_W) \, E(Y) + \lambda^2 x \, E(Y)] .$$

From (13) and (14) we find

$$\int_o^\infty \tilde{H}'(0|x) \, \tilde{R}'(0,x)dx - \int_0^\infty \int_0^\infty y g(dy, t) \int_0^t \lambda E(Y)[E(N_W) + \lambda x]dxdt$$

$$- \lambda E(Y) \int_0^\infty \int_0^\infty y \left[t E(N_W) + \frac{\lambda t^2}{2} \right] g(dy, t)dt$$

$$- \lambda E(Y)E(N_W)E(YS) + \frac{\lambda^2}{2} E(Y)E(YS^2) .$$

Equation (9) is now implied by (10)−(12) and the above.

Three special cases. First, suppose that each arriving item requires 1 unit of space with probability one, i.e., $g(y, t) - 0$ for $y < 1$ and $g(y, t) - g(t)$ if $y \geqslant 1$ for all t. It is then easy to show that $\tilde{Q}(u)$ reduces to the generating function of the number of items in the system, which is the familiar Pollaczek-Khintchine formula.

Next, consider the case where the space occupied by an item is independent of the service time. In this case, $\tilde{B}(u) - \tilde{H}(u|x)$ for all x, so from (6),

$$\tilde{Q}(u) = 1 - \rho + \tilde{B}(u) \int_0^\infty \tilde{R}(u, x)dx \ .$$

This result confirms the fact that the space required by the item in service is independent of that required by the waiting items, conditioned on the server being busy. Moreover, if as before $N(z)$ is the generating function of the number of items in the system, it is possible to show that $\tilde{Q}(u) = N(\tilde{B}(u))$, which is what we expect.

Finally, we note that (6) can be used to calculate the distribution of the unfinished work (virtual waiting time) for the $M/G/1$ queue. In this case, we assume that each item requires an amount of space which is exactly equal to the service time. However, the space required by the item in service depletes steadily as it receives service and is given by

$$Pr\{Y \leqslant y \mid X = x\} = H(y \mid x) = Pr\{S \leqslant x + y \mid S \geqslant x\}$$
$$= [G(x+y) - G(x)]/[1 - G(x)] \ .$$

Further, $g(y, t) = 0$ if $y < t$ and $g(y, t) = g(t)$ if $y \geqslant t$, for all t. Thus, $\tilde{B}(u) = \tilde{G}(u)$. By using (1), (2) and (6), it is now possible to show that

$$\tilde{Q}(u) = \frac{(1-\rho)u}{u - \lambda[1 - \tilde{G}(u)]}$$

is the LST of the unfinished work in the system.

3. A Markov, Finite Memory Model

In the queueing model just described, the item sizes (storage requirements) played a minor role because there was an infinite capacity for storage. In our next two models the dramatic effects of finite capacity will be seen. We also note that certain classical formulas are special cases of the results obtained. In section 5 the results are generalized to networks of storage queues. The development in this and the following two sections is the recent work of Beneš [Ben1].

In the first model items of various sizes arrive in a Poisson stream and are stored in a memory of capacity c. The capacity constraint means that n items of sizes $x_1, ..., x_n$ can be in memory only if $\sum_{i=1}^{n} x_i \leqslant c$. We do not consider the actual arrangement of the items in memory. When n items are in memory, they are being serviced collectively at rate μ_n, i.e., there is probability $\mu_n \Delta t + o(\Delta t)$ that some item will leave in the next interval of time; all are equally likely to be a departing item. The sizes of arriving items are i.i.d. with distribution B and density b. If there is not enough space to accommodate an item at arrival, the item is lost.

Let $p_n(x_1, ..., x_n, t)$ be the density of the probability that n items of sizes $x_1, ..., x_n$ at time t are present, listed in order of arrival. We note that deletion of an item preserves the arrangement by order of arrival, while the arrival of a new item of size x_{n+1} for which there is enough space leads to the addition of a new component at the end of the list. We have

$$
\begin{aligned}
p_n(x_1, ..., x_n, t + \Delta t) &= (1 - \lambda \Delta t - \mu_n \Delta t) \, p_n(x_1, ..., x_n, t) \\
&+ \lambda \Delta t \left[1 - B(c - s_n) \right] p_n(x_1, ..., x_n, t) + \lambda \Delta t \, b(x_n) \, p_{n-1}(x_1, ..., x_n, t) \\
&+ \frac{\mu_{n+1}}{n+1} \sum_{i=1}^{n} \int_0^{c - s_n} p_{n+1}(x_1, ..., x_i, z, x_{i+1}, ..., x_n, t) dz + o(\Delta t),
\end{aligned}
$$

where Δt is positive and sufficiently small, and $s_n = \sum_{i=1}^{n} x_i \leqslant c$. This relation leads to the equilibrium equations

$$
\begin{aligned}
\left[\mu_n + \lambda B(c - s_n) \right] p_n(x_1, ..., x_n) &= \lambda \, p_{n-1}(x_1, ..., x_{n-1}) \, b(x_n) \\
&+ \frac{\mu_{n+1}}{n+1} \sum_{i=0}^{n} \int_0^{c - s_n} p_{n+1}(x_1, ..., x_i, z, x_{i+1}, ..., x_n) dz, \quad s_n \leqslant c.
\end{aligned}
$$

The condition $s_n = \sum_{i=1}^{n} x_i \leqslant c$ that there be sufficient space is symmetric in the x_i, so it is natural to guess that $p_n(x_1, ..., x_n)$ is symmetric, and in fact has the product form

$$
p_n(x_1, ..., x_n) = p_0 \, \frac{\lambda^n}{\prod\limits_{i=1}^{n} \mu_i} \, b(x_1) \, b(x_2) \, \cdots \, b(x_n), \quad \sum_{i=1}^{n} x_i \leqslant c.
$$

That this is a solution can be verified by substitution. The requirement that the probabilities sum to 1 yields

$$1/p_0 = 1 + \sum_{n=1}^{\infty} \frac{\lambda^n}{\prod_{i=1}^{n} \mu_i} \int_{s_n \leqslant c} \cdots \int \prod_{j=1}^{n} b(x_j) dx_1 \cdots dx_n .$$

The probability of loss, defined as

$$P_{loss} = \lim_{t \to \infty} \frac{\text{Number of items lost in } (0, t)}{\text{Number of items arriving in } (0, t)},$$

is given by

$$P_{loss} = p_0 \sum_{n=0}^{\infty} \frac{\lambda^n}{\prod_{i=1}^{n} \mu_i} \int_{s_n \leqslant c} \cdots \int [1 - B(c - s_n)] \prod_{i=1}^{n} b(x_i) dx_1 \cdots dx_n .$$

Examples. For $b(x) = \delta(x - 1)$, c an integer, and $\mu_n = n\mu$ we obtain the classical Erlang-*B* loss formula of telephone traffic theory for the case of exponential holding times with mean $1/\mu$. There, c is the number of trunks, whereas in our model it is the number of unit-sized cells in memory.

For general $c > 0$ and $\mu_n = n\mu$, we model a finite capacity computer memory holding programs being worked on at rate μ, independently of each other. For $\mu_n \equiv \mu$, the stochastic process models the important case of a computer memory of capacity c containing programs with stochastic sizes which are being worked on under a "processor-sharing" discipline; the total processor capability μ is divided equally among the resident programs.

4. Queueing at a Finite Memory

We now describe a modification of the previous model that allows an arriving item to join a queue when there is not enough room for it in memory. Let there be an infinite amount of space for queueing, but only a capacity c in the memory. It is necessary to describe precisely which waiting items are inducted into memory as space becomes available. For simplicity and ease of calculation, let us adopt the protocol whereby queued items are admitted in order of arrival with no passing; this means in particular that a small item is not allowed to move into the memory by passing a larger one ahead of it, when there is room for the former but not the latter. Every item with size greater than c will be rejected, so we will assume that the support of B is in $[0, c]$. The processor requirements for memory, i.e., the residence times of items, are modeled by rates μ_n as before.

Let $p_{mn}(x_1, ..., x_m; y_1, ..., y_n)$ be the density of the equilibrium probability of having m items $x_1, ..., x_m$ in memory being serviced and n others $y_1, ..., y_n$ waiting, all in order of arrival. According to our protocol and capacity constraint, the allowed state vectors $x = (x_1, ..., x_m)$ and $y = (y_1, ..., y_n)$ are such that both

$$\sum_{i=1}^{m} x_i \leqslant c \quad \text{(capacity constraint)}$$

and

$$y_1 > c - \sum_{i=1}^{m} x_i \quad \text{(not enough space for the next item)}.$$

To make maximum use of the symmetries in the problem, it is convenient to define the set A_m, $m \geqslant 2$, to consist of the identity permutation on m objects together with the permutations $\{\sigma_i; i=1, ..., m-1\}$, such that the action of σ_i on a vector $x \in R^m$ is to insert x_1 between x_{i+1} and x_{i+2}, i.e.,

$$\sigma_i x = (x_2, x_3, ..., x_{i+1}, x_1, x_{i+2}, ..., x_m) .$$

Of course, $x_{i+2}, ..., x_m$ does not appear when $i + 1 = m$. Using A_m we define the cyclic average, $\bar{p}_{mn}$, as

$$\bar{p}_{mn}(x;y) = \frac{1}{m} \sum_{\sigma \in A_m} p_{mn}(\sigma x; y), \quad m = |A_m| .$$

For FCFS queueing, the stationary equations are derived as before. With $s_j = \sum_{i=1}^{j} x_i$, $s_0 = 0$,

$$\lambda p_0 = \mu_1 \int_0^c p_{10}(x_1) dx_1 ,$$

$$(\mu_m + \lambda) p_{m0} = \lambda p_{m-1,0} b(x_m) + \mu_{m+1} \int_0^{c-s_m} \bar{p}_{m+1,0}(z, x_1, ..., x_m) dz$$

$$+ \sum_{j=0}^{m-1} \mu_{j+1} \int_{c-s_{j+1}}^{c-s_j} \bar{p}_{j+1, m-j}(z, x_1, ..., x_j; x_{j+1}, ..., x_m) dz, \quad m \geqslant 1 ,$$

$$(\mu_m + \lambda) p_{mn} = \lambda p_{m,n-1} b(y_n) + \mu_{m+1} \int_0^{c-s_m} \bar{p}_{m+1, n}(z, x_j, ..., x_m; y) dz$$

$$+ \sum_{j=1}^{m-1} \mu_{j+1} \int_{c-s_{j+1}}^{c-s_j} \bar{p}_{j+1, m-j-n} (z, x_1, ..., x_j; x_{j+1}, ..., x_m, y)dz, m, n > 0 ,$$

where p_0 is found from the normalization condition.

The significance of the cyclic averages $\bar{p}_{mn}$ is that if $z, x_1, ..., x_j$ are in memory, then all are equally likely to be finished by the processor in the next Δt time units, so that the one which actually leaves, z, could be in any of $j + 1$ positions in the arrival sequence. It is also important to note that the size of an item that has to wait is exactly the amount of memory it will occupy when it finally gets in. Thus, the model obeys the conservation principle described earlier. Of course, this property involves no cost here, because the waiting size has not yet really interacted with the system; see, however, the network model in the next section, which satisfies the conservation principle even with interaction. We now present two examples worked out by Beneš [Ben1].

Example 1. For c an integer and $b(x) = \delta(x-1)$, the choice $\mu_m = m\mu$ for both $1 \leqslant m \leqslant c$ and $m > c$ suggests a relation to the M/M/c formula, and a corresponding solution. We try

(15)
$$p_{m0}(x; y) = p_0 \frac{\rho^m}{m!} \prod_{i=1}^{m} \delta(x_i - 1) , \quad m \leqslant c ,$$

$$p_{cn}(x; y) = p_0 \frac{\rho^c}{c!} \left[\frac{\rho}{c} \right]^n \prod_{i=1}^{c} \delta(x_i - 1) \prod_{j=1}^{n} \delta(y_j - 1) ,$$

where $\rho = \lambda/\mu$. The stationary equations simplify to

$$(m\mu + \lambda)p_{m0} = \lambda p_{m-1,0} \delta(x_m - 1) + (m+1)\mu \int_0^{c-m} \bar{p}_{m+1,0} (z, x_1, ..., x_m)dz ,$$

$$(c\mu + \lambda)p_{cn} = \lambda p_{c,n-1} \delta(y_n - 1) + c\mu \int_0^1 \bar{p}_{c,n+1} (z, x_1, ..., x_{c-1}; x_c, y)dz$$

and substitution shows that (15) is the desired nonnegative solution. Integrating out all of the delta functions, we obtain the classical result for the M/M/c queue.

Example 2. The important processor-sharing case $\mu_m \equiv \mu > 0$ is readily solvable. In order to give a nonnegative solution we guess the product form

(16) $p_{mn}(x;y) = p_0 \rho^{m+n} \prod\limits_{i=1}^{m} b(x_i) \prod\limits_{j=1}^{n} b(y_j), y_1 > c - \sum\limits_{i=1}^{m} x_i$.

It is clear that

$$\mu p_{mn}(x;y) = \lambda p_{m,n-1}(x;y_1, ..., y_{n-1}) b(y_n)$$
$$\mu p_{m0}(x) = \lambda p_{m-1,0}(x_1, ..., x_{m-1}) b(x_m)$$

and also that $\bar{p}_{mn} = p_{mn}$. Moreover,

$$\mu \left[\int_0^{c-s_m} p_{m+1,n}(z,x;y)dz + \int_{c-s_m}^{c-s_{m-1}} \bar{p}_{m,n+1}(z,x_1, ..., x_{m-1}; x_m y)dz \right.$$

$$\left. + \cdots + \int_{c-x_1}^{c} \bar{p}_{1,m+n}(z;x,y)dz \right] =$$

$$\mu p_0 \rho^{m+n+1} \sum_{i=1}^{m} b(x_i) \prod_{i=1}^{n} b(y_j) \left[\int_0^{c-s_m} + \int_{c-s_m}^{c-s_{m-1}} + \cdots + \int_{c-x_1}^{c} \right] b(z)dz$$

$$= \lambda p_{mn}(x;y) ,$$

since the sum of integrals on the right is exactly $B(c) = 1$. Thus, the function in (16) is a positive solution.

5. Networks of Finite Memories and Shared Processors

The analysis of the preceding two sections can be generalized to some extent to the computer/communication networks modeled by the Jackson network [Ja] and its generalizations. We shall see that the product-form solutions for stationary state probabilities are preserved when the state-variable is extended to include storage states. We shall content ourselves with Beneš development [Ben1] of the balance equations; computational problems remain an open area of research.

Let L be a given set of nodes α. At each node there is a system consisting of a queue, a memory and a processor. At node α the memory has capacity c_α, and the processor has rate μ_α. Items that finish at one node may leave the system or move on to another node in accordance with a substochastic matrix $P = ((p_{\alpha\beta}))$, with $p_{\alpha\alpha} = 0$ and

$$p_{\alpha\beta} = Pr \{\text{an item leaving } \alpha \text{ goes to } \beta \text{ for more service}\},$$
$$p_\alpha = 1 - \sum_{\beta \epsilon L} p_{\alpha\beta} = Pr \{\text{an item leaving } \alpha \text{ exits the system}\}.$$

We shall insist on the conservation of item size; once a size is drawn according to B, it remains the same until the item leaves the system. The processor-sharing discipline is in force at each node.

We use the state variable

$$\xi = \left\{ x_1^\alpha, ..., x_m^\alpha; y_1^\alpha, ..., y_n^\alpha, \ m = m(\xi,\alpha), \ n = n(\xi, \alpha), \ \alpha \epsilon L \right\}$$

to mean that in the order of their arrival at node α, items of size $x_1^\alpha, ..., x_m^\alpha$ are in memory at α, while $y_1^\alpha, ..., y_n^\alpha$ are waiting to get in. The integers m and n are of course also part of the state. Items are entered into memory in strict order of arrival. The possible states are subject to the memory size constraint

$$\sum_{j=1}^{m(\alpha,\xi)} x_j^\alpha \leqslant c_\alpha, \quad \alpha \epsilon L$$

and to the sufficient-space condition

$$c_\alpha - \sum_{j=1}^{m(\alpha,\xi)} x_j^\alpha < y_1^\alpha, \quad \alpha \epsilon L .$$

Inevitably, the exact content in our state variable leads to combinatorial problems in describing the many possible transitions that can result in state ξ:

(1) ξ arises from a new item's arriving and entering memory at node α. In this case $n = n(\alpha,\xi) = 0$, the arrival's size was x_m^α, and

$$c_\alpha - \sum_{j=1}^{m-1} x_j^\alpha \geqslant x_m^\alpha$$

i.e., there was space available. The state preceding ξ was $\xi_{\alpha x}$, defined as ξ with x^α replaced by $x_1^\alpha, ..., x_{m-1}^\alpha$ (last component removed).

(2) ξ arises from a new item's arriving at α and joining the queue there, either because it found no queue but did not fit in memory, or it found at least one item waiting. In this case $n \geqslant 1$, and y_n^α is the arrival's size. If it found no queue but could not get into memory, then $n = 1$ and

$$c_\alpha - \sum_{j=1}^m x_j^\alpha < y_n^\alpha .$$

The state preceding ξ is $\xi_{\alpha y}$, defined as ξ with y^α replaced by $y_1^\alpha, ..., y_{n-1}^\alpha$ (last component removed). Obviously, if $n = 1$, y^α is not replaced by anything, it is simply removed. Similarly, in case (1), if $m = 1$, then $x^\alpha = x_1^\alpha$ is simply removed.

(3) ξ results from an item's ending its service at a node $\beta \neq \alpha$, moving to α, and entering the memory there. The item's size was x_m^α and there was no queue at α, i.e., $n(\alpha,\xi) = 0$. Some number i, $0 \leqslant i \leqslant m(\beta,\xi)$, of items with sizes $x_{m-i+1}^\beta, ..., x_m^\beta$ were able to move into β's memory as a result of x_m^α's departure, where i is the integer such that

$$\sum_{j=m-i+1}^m x_j^\beta \leqslant c_\beta - \sum_{j=1}^{m-i} x_j^\beta + x_m^\alpha < \sum_{j=m-i+1}^m x_j^\beta + y_i^\beta .$$

Together, these inequalities indicate that there was space for the items $x_{m-i+1}^\beta, ..., x_m^\beta$ waiting when x_m^α left, but that the next one waiting at β, viz y_i^β, did not fit. Note that the occurrences of m above are really $m(\beta,\xi)$, except the lower index on x_m^α, which is $m(\alpha,\xi)$.

In addition, in the arrival sequence at β which is part of the state ξ, the transferring item of size x_m^α could have been in any one of $m-i+1$ positions, as shown below.

$$x_m^\alpha, x_1^\beta, ..., x_{m-i}^\beta$$

$$x_1^\beta, x_m^\alpha, x_2^\beta, ..., x_{m-i}^\beta$$

$$\vdots \qquad\qquad m = \begin{cases} m(\beta) \text{ in } x^\beta \\ \\ m(\alpha) \text{ in } x^\alpha \end{cases}$$

$$x_1^\beta, ..., x_{m-i}^\beta, x_m^\alpha$$

These possibilities can be indexed as before by an $(m-i+1)$-permutation in A_{m-i+1}. Accordingly, the possible states preceding ξ are denoted $\xi_{\beta\alpha x i \sigma}$, $i=0, ..., m$ (β, ξ), $\sigma \epsilon A_{m-i+1}$, and defined as ξ with

$$x^\beta \text{ replaced by } \sigma\left[x^\alpha_m, x^\beta_1, ..., x^\beta_{m-i}\right], \quad m = m(\beta) ,$$
$$x^\alpha \text{ replaced by } x^\alpha_1, ..., x^\alpha_{m-1}, \quad m = m(\alpha) ,$$
$$y^\beta \text{ replaced by } x^\beta_{m-i+1}, ..., x^\beta_m, y^\beta, \quad m = m(\beta) .$$

A value $i = 0$ means that the additional space vacated by x^α_m was not enough to accommodate y^β_1; i.e.,

$$y^\beta_1 > c_\beta - \sum_{j=1}^{m(\beta)-1} x^\beta_j .$$

(4) ξ is produced by an item's finishing service at a node $\beta \neq \alpha$, transferring to α and either finding no queue at α but insufficient space to enter memory, or finding a queue. Then $n(\alpha, \xi) > 0$, and the size of the transferring item is y^α_n. Again, as in case (3), some number i, $0 \leqslant i \leqslant m(\beta, \xi)$ of items with sizes $x^\beta_{m-i+1}, ..., x^\beta_m$ were able to move into β's memory. Prior to transfer, y^α_n occupied a position ahead of, between or after the components $x^\beta_1, ..., x^\beta_{m-i}$. Again, this position can be indexed by $\sigma \epsilon A_{m-i+1}$. The states preceding ξ are then of the form $\xi_{\beta\alpha y i \sigma}$, $i=0, ..., m(\beta, \xi)$, $\sigma \epsilon A_{m-i+1}$ and defined as ξ with

$$x^\beta \text{ replaced by } \sigma\left[y^\alpha_n, x^\beta_1, ..., x^\beta_{m-i}\right], \quad m = m(\beta, \xi) ,$$
$$y^\alpha \text{ replaced by } y^\alpha_1, ..., y^\alpha_{n-1}, \quad n = n(\alpha, \xi) ,$$
$$y^\beta \text{ replaced by } x^\beta_{m-i+1}, ..., x^\beta_m, y^\beta ,$$

where i is such that

$$\sum_{j=m-i+1}^m x^\beta_j \leqslant c_\beta - \sum_{j=1}^m x^\beta_j + y^\alpha_n < \sum_{j=m-i+1}^m x^\beta_j + y^\beta_1 .$$

In addition, either $n(\alpha, \xi) > 1$, or $n(\alpha, \xi) = 1$ and

$$c_\alpha - \sum_{j=1}^{m(\alpha,\xi)} x^\alpha_j < y^\alpha_n ,$$

since y^α_n had to wait in queue.

(5) ξ arises because an item ended its stay in memory at node α and left the system. The size, z, of the departed item is not identifiable from the state ξ, but it could have been no larger than $c_\alpha - \sum\limits_{j=1}^{m} x_j^\alpha$. For some i, items with sizes $x_{m-i+1}^\alpha, ..., x_m^\alpha$ were able to move into α's memory as a result of the additional z units of space, and the departed item could have been (in the arrival order at α) at any one of the $m-i+1$ positions among $x_1^\alpha, ..., x_{m-i}^\alpha$. Thus, the possible states preceding ξ under this type of transition can be denoted $\xi^{\alpha z i \sigma}$ and defined as ξ with

$$x^\alpha \text{ replaced by } \sigma(z, x_1^\alpha, ..., x_{m-i}^\alpha) ,$$
$$y^\alpha \text{ replaced by } x_{m-i+1}^\alpha, ..., x_m^\alpha, y^\alpha ,$$

where i is an integer such that

$$0 \leqslant c_\alpha - \sum_{j=1}^{m-i} x_j^\alpha - z < x_{m-i+1}^\alpha .$$

If

$$z \leqslant c_\alpha - \sum_{j=1}^{n(\alpha,\xi)} x_j^\alpha < y_1,$$

then $i = 0$ and no items moved into memory with the departure of size z.

The five types of transitions just described exhaustively determine the set of possible states preceding state ξ. The notation we have introduced eases the difficulty of writing the equilibrium equations, as can be seen in the following basic relation. We have

(17)
$$\sum_{\alpha \in L} (\mu_\alpha \chi (m(\alpha) > 0) + \lambda_\alpha) p(\xi) =$$

$$\sum_{\substack{\alpha \in L \\ m(\alpha) > 0}} \lambda_\alpha \left[p(\xi_{\alpha x}) \, b(x_m^\alpha) \, 1_{n(\alpha) = 0} + p(\xi_{\alpha y}) \, b(y_n^\alpha) \right]$$

$$\left[\chi(n(\alpha) > 1) + \chi(n(\alpha) = 1) \, \chi(y_n^\alpha > c_\alpha - s_m^\alpha) \right]$$

$$+ \sum_{\substack{\alpha,\beta\epsilon L \\ m(\alpha)>0}} \mu_\beta P_{\beta\alpha} \sum_{i=0}^{m(\beta)} \frac{1}{m(\beta)-i+1} \left[p\left(\xi_{\beta\alpha x i\sigma}\right) \chi(n(\alpha)=0)\right.$$

$$\left. + p\left(\xi_{\beta\alpha y i\sigma}\right) \chi(n(\alpha)>0)\right]$$

$$+ \sum_{\alpha\epsilon L} \mu_\alpha P_\alpha \sum_{i=0}^{m(\alpha)} \frac{1}{m(\alpha)-i+1} \sum_{\sigma\epsilon A_{m-i+1}} \int_{c_\alpha-s^\alpha_{m-i+1}}^{c_\alpha-s^\alpha_{m-i}} p\left(\xi^{\alpha z i\sigma}\right) dz \; ,$$

where both s^α_0 and $c_\alpha - s^\alpha_{m+1} = c_\alpha - \sum_{j=1}^{m+1} x^\alpha_j$ are interpreted as 0, and $\chi(A)$ is the indicator function of event A. The sums on i are over the $m-i+1$ possible positions of the arrival sequence which the transferring or departing item could occupy, and the factors $(m-i+1)^{-1}$ reflect the fact that processor sharing makes all $m-i+1$ items in memory in the preceding state equally likely to complete in the next Δt time units.

These equations can be solved by combining the ideas on exact content in section 3 with the Jackson throughput equation, which in the present case has the form

$$(18) \qquad \theta_\alpha = \sum_{\beta\epsilon L} \theta_\beta P_{\beta\alpha} + \lambda_\alpha \; , \qquad \alpha\epsilon L \; .$$

The solution vector, θ, is to be interpreted as the effective arrival rate, taking into account all of the paths that a given item can follow. Indeed, λ_α is a "source" term, and the sum collects all of the "reflection" terms aimed at α. If $\theta < \mu$ component by component, then an integrable equilibrium density $p(\xi)$ exists and is given by

$$(19) \qquad p(\xi) = p(0) \prod_{\alpha\epsilon L} (\rho_\alpha)^{m+n} \prod_{j=1}^{m} b(x^\alpha_j) \prod_{i=1}^{n} b(y^\alpha_i) \; ,$$

where of course $m = m(\alpha,\xi)$, $n = n(\alpha,\xi)$, and $\rho_\alpha = \theta_\alpha/\mu_\alpha$. It is not hard to see that the product form (19) is indeed a solution of the stationary equations; one has only to do the bookkeeping, using the Jackson equation (18) and

$$(20) \qquad \sum_{\alpha\epsilon L} \lambda_\alpha = \sum_{\alpha\epsilon L} P_\alpha \theta_\alpha \; .$$

6. A Model of Fragmentation

To this point the criterion for admitting new items into storage has been based on whether the total of unused space in memory is sufficient. In this section we turn to the analysis of systems in which this criterion must be generalized to account for different *patterns* of unused space, i.e. the fragmentation of storage produced by the scattering of occupied and unoccupied locations throughout memory. Background for these dynamic storage allocation problems is provided in section 1.2 (see also [Ben2]). To fix ideas here, we begin with the following discretized mathematical model.

Consider an unbounded, linear memory represented simply by the set, $\{1,2,...\}$, of possible locations or addresses. We have a Poisson arrival at rate $\lambda > 0$ of items to be stored in memory. Each item has a size given by a positive integer. At its time of arrival an item of size ℓ must be placed into some subset of ℓ *consecutive* unoccupied locations in memory, the set chosen being determined by an allocation algorithm or rule. On being placed into storage, each item commences a residence time; at the end of this period the item departs from storage, leaving the locations it occupied available for subsequent arrivals. The memory clearly becomes fragmented over time, i.e. the unused space exists in a sequence of "holes" alternating with regions currently occupied by items.

More precisely, let $S_t \subset \{1,2,...\}$ denote the state of memory at time t; an integer is in S_t if and only if the corresponding location is occupied at time t. For simplicity we assume that item residence times are independent, exponential random variables, each with parameter $\mu > 0$. Also, we concentrate on the First-Fit (FF) allocation rule: When an item of size ℓ arrives, it is stored in the interval of unoccupied locations $\{i, i+1,...,i+\ell-1\}$ for which i is minimum.

In general, of course, $M = \{S_t, t \geqslant 0\}$ is not a Markov process, since for this purpose a state must also specify the occupied locations belonging to each item currently in memory. However, the system can be analyzed under the assumption that all items have the same size. In this special case, M is indeed a continuous time Markov process. Normalizing the common item size to 1 we see that transitions occur at rate $\mu > 0$ from any non-empty state S to *each* of the $|S|$ subsets of S obtained by deleting one location from S, and at rate λ from any S to the union of S and the smallest numbered location not in S. Because S diminishes at rate $|S|\mu$, it is easy to see that for any λ and μ, if $|S_0| < \infty$ then $|S_t| < \infty$ for all $t > 0$ with probability 1 and that M is a Markov chain with a stationary distribution on the set of finite subsets of $\{1,2,...\}$. Even with the simplification to equal item sizes, it does not seem possible to obtain the stationary distribution of S. However, we can find this distribution for both $|S|$ and $\max(S)$, the maximum element of S. Note that $\max(S)$ is simply the total occupancy of memory, and that $\max(S)-|S|$ is the wasted space (the total size of the holes in S).

The analysis of $\max(S)$ follows that in [CofKS1] and is the primary goal of this section. Later, we investigate the asymptotic behavior of the expected amount of wasted space. Numerical results for non-degenerate item size distributions are cited from [CofKS2] as support for a conjecture that the FF rule is asymptotically optimal in a strong sense.

The Analysis of **max(S)**. For simplicity we continue to normalize the common item size to 1. We note immediately that the analysis of the total number of occupied locations, $|S|$, is well-known in the guise of the $M/M/\infty$ (infinite server) queue. In particular, letting $\pi_k = \lim_{t \to \infty} Pr\{|S_t| = k\}$, $k = 0,1,...$, denote the stationary distribution, we have the following standard results [Fel1] in terms of $\rho = \lambda/\mu$:

(21)
$$\pi_k = \frac{\rho^k}{k!} e^{-\rho}, \quad k = 0,1,...,$$

and

(22)
$$E(|S|) = \sum_{k \geq 0} k\pi_k = \rho .$$

It is not difficult to see that $\max(S)$ is *not* a Markov process. We get around this difficulty by first analyzing an embedded two-dimensional Markov process and then applying a special technique with generating functions. The embedded Markov process with parameter m is $(|S_t \cap \{1,2,...,m\}|, |S_t \cap \{m+1,m+2,...\}|)$, which is Markovian on states of integer pairs (k,r), $0 \leq k \leq m$, $0 \leq r$. The stationary distribution, for fixed $m \geq 0$, is denoted

(23) $\pi(k,r) = \pi(k,r;m)$

$\qquad = Pr\{|S \cap \{1,...,m\}| = k , \quad |S \cap \{m+1,m+2,...\}| = r\} .$

Note that $Pr\{\max(S) \leq m\} = Pr\{|S \cup \{m+1,m+2,...\}| = 0\}$, and that $\pi(k,r)$ satisfies the equations

(24) $\pi(k,r)(\lambda+(k+r)\mu) = \pi(k-1,r)\lambda + \pi(k+1,r)(k+1)\mu + \pi(k,r+1)(r+1)\mu$

for $0 \leq k < m$ and $r \geq 0$, where $\pi(-1,r) = 0$, and

(25) $\pi(m,r)(\lambda+(m+r)\mu) = \pi(m-1,r)\lambda + \pi(m,r-1)\lambda + \pi(m,r+1)(r+1)\mu$

for $r \geq 0$, where $\pi(m,-1) = 0$. Introducing the generating function

(26) $$F(x,y) = \sum_{k=0}^{m} \sum_{r=0}^{\infty} \pi(k,r)x^k y^r ,$$

and $\rho = \lambda/\mu$, we obtain from (23) and (24)

(27) $$\rho(1-x)F(x,y) + (x-1)\frac{\partial}{\partial x}F(x,y) + (y-1)\frac{\partial}{\partial y}F(x,y)$$
$$= \rho x^m (y-x)f_m(y)$$

where

(28) $$f_m(y) = \sum_{r=0}^{\infty} \pi(m,r)y^r = \frac{1}{m!}\frac{\partial^m}{\partial x^m}F(x,y) .$$

In view of the fact that

(29) $$Pr\{\max(S) \leq m\} = \sum_{k=0}^{m} \pi(k,0) = F(1,0) ,$$

we now want to find $F(1,0)$. The difficulty in solving for F from (27) and (28) is that it is a partial differential equation of degree m. However, by alternately setting $x = 1$ and $y = 1$ and repeatedly differentiating, we can reduce (27) to an infinite series of ordinary differential equations which we can solve.

First, setting $x = 1$ in (27) and noting that $F(1,1) = 1$, we obtain

(30) $$F(1,y) = 1 - \rho \int_{y}^{1} f_m(u)du .$$

Next, setting $y = 1$ in (27) we obtain

$$(31) \qquad F(x,1) - \frac{1}{\rho} \frac{\partial}{\partial x} F(x,1) = x^m f_m(1) ,$$

and solving this (ordinary) differential equation, with $F(1,1) = 1$, we find

$$(32) \qquad F(x,1) = \rho e^{\rho x} \int_x^\infty e^{-\rho u} u^m du \, f_m(1) .$$

Now, for each $n \geq 1$, differentiate (27) n times with respect to y and then set $y = 1$, to obtain

$$(33) \quad \rho(1-x) \frac{\partial^n}{\partial y^n} F(x,1) + (x-1) \frac{\partial}{\partial x} [\frac{\partial^n}{\partial y^n} F(x,1)] + n \frac{\partial^n}{\partial y^n} F(x,1) =$$
$$\rho x^m (1-x) f_m^{(n)}(1) + n \, \rho x^m f_m^{(n-1)}(1)$$

where

$$(34) \qquad f_m^{(n)}(y) = \frac{\partial^n}{\partial y^n} f_m(y) , \quad n = 0,1,2,\dots .$$

Solving (33) for $\frac{\partial^n}{\partial y^n} F(x,1)$ we obtain

$$(35) \qquad \frac{\partial^n}{\partial y^n} F(x,1) = - \frac{1}{(x-1)^n} \rho e^{\rho x} \int_x^\infty du \, e^{-\rho u} (u-1)^{n-1}$$
$$[n^m (1-u) f_m^{(n)}(1) + n u^m f_m^{(n-1)}(1)] , \quad n = 1,2,\dots .$$

Since $F(x,y)$ is a polynomial in x for any y, the right side of (35) must be finite at $x = 1$ and hence the integral in (35) must vanish at $x = 1$, namely,

$$(36) \int_1^\infty du \, e^{-\rho u} (u-1)^{n-1} [u^m (1-u) f_m^{(n)}(1) + n u^m f_m (n-1)(1)] = 0 , \quad n \geq 1 .$$

Rewriting (36) as

$$(37) \qquad \frac{f_m^{(n)}(1)}{n!} \int_1^\infty du \, e^{-\rho u} (u-1)^n u^m$$
$$- \frac{f_m^{(n-1)}(1)}{(n-1)!} \int_1^\infty du \, e^{-\rho u} (u-1)^{n-1} u^m , \quad n \geq 1 ,$$

we observe that the left side of (37) must be the same for all $n = 0,1,2,...$. Setting $n = 1$ in (37), then using (32) and the fact that $F(1,1) = 1$, we have

$$(38) \qquad \frac{f_m^{(n)}(1)}{n!} \int_1^\infty du \ e^{-\rho u} (u-1)^n u^m = \frac{f_m^{(0)}(1)}{0!} \int_1^\infty du \ e^{-\rho u} u = \frac{e^{-\rho}}{\rho} .$$

Now, using Taylor's expansion about $y = 1$,

$$(39) \qquad f_m(y) = \sum_{n=0}^\infty \frac{f_m^{(n)}(1)(y-1)^n}{n!} .$$

Since the coefficients $\pi(m,r)$ in (38) satisfy $\pi(m,r) \leqslant P(|S| \geqslant m+r) = O(1/r!)$ by (21), f_m is an entire function and the validity of (39) follows. Using (38) and (39), we have

$$(40) \qquad f_m(y) = \frac{e^{-\rho}}{\rho} \sum_{n=0}^\infty (y-1)^n \frac{1}{\int_1^\infty du \ e^{-\rho u} (u-1)^n u^m} .$$

Then, from (30)

$$(41) \qquad F(1,y) = 1 - e^{-\rho} \sum_{n=0}^\infty \frac{(y-1)^{n+1}}{(n+1)} \frac{1}{\int_1^\infty du \ e^{-\rho u} (u-1)^n u^m} .$$

Setting $y = 0$ we have from (29) the main result,

$$(42) \ Pr\{\max(S) > m\} = \sum_{n=0}^\infty \frac{(-1)^n}{n+1} \frac{1}{\sum_{k=0}^m \binom{m}{k} (n+k)!/\rho^{n+k+1}} , \qquad m = 0,1,... .$$

Note that for $m = 0$, $Pr\{\max(S) > 0\} = 1 - P(S = \varnothing) = 1 - Pr\{|S| = 0\} = 1 - e^{-\rho}$, from (21). As can be seen, (42) agrees with this result.

We observe finally that the joint distribution of $\max(S)$ and $|S|$ can be obtained from $F(x,y)$, which is determined from (35) and (38) explicitly. Indeed,

$$Pr\{\max(S) \leqslant m , \ |S| \leqslant k\} = \sum_{j=0}^k \pi(j,0;m)$$

and $\pi(j,0;m) = \pi(j,0) = \partial^j/\partial x^j \ F(0,0)$, which can be obtained from $F(x,y)$.

As indicated in [CofKS1] the resulting expression for the joint distribution of $\max(S)$ and $|S|$ does not appear to have a simple form, so we do not pursue this problem further.

The Rate of Wasted Space. The rate of wasted space, defined as

$$W(\rho) = \frac{E(\max(S)) - E(|S|)}{E(|S|)} \, ,$$

can be calculated as a function of $1/E(|S|) = 1/\rho$ based directly on (42). Numerical computations in [CofKS1] show that $W(\rho) \to 0$ monotonically as $\rho \to 0$ and $\rho \to \infty$, and that a maximum clearly less than $1/3$ exists at $\rho = E(|S|) \cong 40$. From the strictly monotone behavior on both sides of the maximum the bound $E(\max(S)) \leqslant (4/3) E(|S|)$ was observed.

In [CofKS2] it is conjectured that the *FF* rule is asymptotically optimal, i.e. $W(\rho) \to 0$ as $\rho \to \infty$, for *any* distribution of item sizes (with finite mean). Rather convincing numerical evidence supporting this conjecture will be cited at the end of this section.

It may be surprising at first that the rate of wasted space approaches 0 as $\rho \to \infty$, since the intuitive fact that the expected number of holes grows indefinitely with increasing ρ is not hard to prove. Because of its importance, and because direct calculation is quite awkward for large m and ρ in (42), we next prove analytically that $W(\rho) \to 0$ as $\rho \to \infty$ for the case of a single item size. To underscore the unexpected nature of the general result we then prove that for a general distribution of item sizes the expected number of holes increases linearly with $E(|S|)$.

First, the following bounds will be useful:

$$(43) \qquad 1 - \gamma_m \leqslant Pr\{\max(S) > m\} \leqslant \rho \, \frac{\gamma_m - \gamma_{m-1}}{\gamma_m} \, ,$$

where

$$(44) \qquad \gamma_m = e^{-\rho} \sum_{k=0}^{m} \frac{\rho^k}{k!} \, , \quad m = 0, 1, \dots .$$

Observe first that $\max(S) \geqslant |S|$, so that $Pr\{\max(S) > m\} \geqslant Pr\{|S| > m\}$; the first inequality thus follows from (21) and (44). For the second, note that

$$(45) \qquad Pr\{\max(S) > m\} \leqslant E\left(|S \cap \{m+1, m+2, ...\}|\right)$$
$$= E\left(|S|\right) - E\left(|S \cap \{1, 2, ..., m\}|\right) .$$

Now observe that $|S_t \cap \{1, ..., m\}|$ is an embedded Markov process on states $0, 1, ..., m$, and that its stationary distribution can be easily obtained as

$$(46) \qquad Pr\{|S \cap \{1, ..., m\}| = k\} = \frac{\rho^k}{k!} \frac{1}{\displaystyle\sum_{j=0}^{m} \frac{\rho^j}{j!}}, \qquad k = 0, 1, ..., m .$$

Using (46) in (45) gives the second inequality in (43).

We can use (43) to show that the fraction of wasted space is asympotically negligible for $\rho \rightarrow \infty$. First note that, for any integer K,

$$(47) \qquad E[\max(S)] = \sum_{m=0}^{\infty} Pr\{\max(S) > m\}$$
$$\leqslant K + \sum_{m=K}^{\infty} Pr\{\max(S) > m\} .$$

Choosing $K = (1+\epsilon)\rho$, using (47) and noting that γ_m approaches 1 from below, we have

$$(48) \qquad \frac{E(\max(S))}{\rho} \leqslant 1 + \epsilon + \frac{1 - \gamma_K}{\gamma_K} .$$

Since for every $\epsilon > 0$, $1 - \gamma_K = o(1)$ as $\rho \rightarrow \infty$, it follows from (48) that

$$(49) \qquad \overline{\lim_{\rho \rightarrow \infty}} \frac{E(\max(S))}{\rho} \leqslant 1 .$$

On the other hand, $\max(S) \geqslant |S|$, and therefore $E(\max(S)) \geqslant E(|S|) = \rho$. Hence

$$(50) \qquad E(\max(S) - |S|) = o(E(|S|)) \quad \text{as} \quad \rho \rightarrow \infty .$$

Thus, as $\rho \rightarrow \infty$ the total size of the holes becomes negligible relative to the number of occupied locations.

The method of (47) can be used to tighten the bound of (48) on $E(\max(S))$ by applying (47) for $K = \sqrt{\rho \log \rho}$ instead of $K = (1+\epsilon)\rho$. In fact, it can be shown that [CofKS1]

$$(51) \qquad E(\max(S)-|S|) \leqslant c\sqrt{\rho \log \rho}$$

for some constant c.

Let us now consider the general case, where we have an arbitrary distribution F of item sizes. Also, for convenience we normalize the arrival rate, $\lambda = 1$. We prove that if F is continuous and if h denotes the number of holes in the stationary regime, then $E(h) \sim \rho/2$ as $\rho \to \infty$. This is an instance of Knuth's [Knu] "fifty-percent rule," which states that in equilibrium the expected number of holes is approximately one half the expected number of items present.

Consider a random set, S, of items with the stationary distribution. Following Knuth [Knu] let N_0 denote the number of items of S which when removed cause h to increase by one (these are called interior items). Similarly, let N_1 be the number of items of S which when removed do not change h (called boundary items), and let N_2 be the number of items of S which when removed decrease h by one (called isolated items). Clearly the total number of items N in S is

$$(52) \qquad N = N_0+N_1+N_2 .$$

It is easy to see that the number of holes in S satisfies

$$(53) \qquad 2h = N_1+2N_2+\xi , \quad |\xi| \leqslant 2 ,$$

since every hole (except the one beginning at zero if there is one) has a left and right hand neighboring item. Indeed, no interior item borders a hole, each isolated item is counted at most twice, and each boundary item is counted at most once. Thus, we have $2(h-1) \leqslant N_1+2N_2$. Also, every boundary item (except the extreme right one if it has an item to its left) borders a hole, and every isolated item borders two holes. Therefore, $N_1+2N_2 \leqslant 2h+1$ and (53) follows, since every hole is counted at most twice.

Now start the storage process S_t in its stationary state $S_0 = S$, and let it run for a small time interval of length $t > 0$. Then, with $\chi(A)$ denoting the indicator of event A,

(54) $\qquad E(h) = E(h(t)) = E(E[h(t)|S])$

$\qquad\qquad = E(E[h(t)\chi(\text{nothing happens in } (0,t))|S])$

$\qquad\qquad + E(E[h(t)\chi(\text{an arrival occurs in } (0,t))|S])$

$\qquad\qquad + E(E[h(t)\chi(\text{an interior item departs in } (0,t))|S])$

$\qquad\qquad + E(E[h(t)\chi(\text{a boundary item departs in } (0,t))|S])$

$\qquad\qquad + E(E[h(t)\chi(\text{an isolated item departs in } (0,t))|S])$

$\qquad\qquad + o(t) \qquad \text{as } t \to 0 ,$

where the last term follows from the fact that the probability of more than one of the events indicated in (54) is $o(t)$ as $t \to 0$. Then we have for the corresponding terms with $\alpha = Pr\{\text{an arriving item exactly fills a hole of } S\}$,

(55) $\qquad E(h) = E(h)(1-t-N\mu t)+E(h)t-\alpha t$

$\qquad\qquad + E(h+1)N_0\mu t+E(h)N_1\mu t + E(h-1)N_2\mu t+o(t) ,$

since the probability of an arrival is $t+o(t)$ and a departure of type N_i is $N_i\mu t+o(t)$ as $t \to 0$. Canceling terms in (55) using (52) and dividing by t and letting $t \to 0$ gives immediately

(56) $$E(N_2) = E(N_0) - \frac{\alpha}{\mu} .$$

Now from (53) we have, taking expectations and using (56) and (52)

$$2E(h) = E(N_1+N_2+N_2)+E(\xi)$$

$$= E(N_0+N_1+N_2) - \frac{\alpha}{\mu} + E(\xi)$$

$$= E(N) - \frac{\alpha}{\mu} + E(\xi) = \frac{1-\alpha}{\mu} + E(\xi)$$

For the case of continuous F, $\alpha = 0$. Thus, since $|\xi| \leqslant 2$, we have as claimed $E(h) \sim 1/(2\mu) = \rho/2$ as $\rho \to \infty$.

For the case of a single item size, where $W(\rho) \to 0$ as $\rho \to \infty$ has been proved, it now follows that $\alpha = \alpha(\rho) \to 1$ as $\rho \to \infty$; i.e. the arriving item exactly fills its space with probability approaching 1 as $\rho \to \infty$. This is consistent with intuition, since as $\rho \to \infty$ there are many items and the first hole seems likely (and is now proved) to be of the size of one item.

That there are $o(\rho)$ holes in the remaining cases is surprising. The only way that $W(\rho) \to 0$ and $E(h) \to \infty$ as $\rho \to \infty$ can both hold is if the set of holes in the stationary regime must be comprised of very many small ones and only a few, i.e. $o(\rho)$, which exceed δ for any $\delta > 0$.

In [CofKS2] Monte Carlo experiments were conducted in order to test the conjectured asymptotic optimality of FF. Since intuitively uniform distributions were expected to be a severe test, the majority of the results apply to this easily simulated case. Based on sequences of 10^5 arrivals, the conjecture was supported by the results. More precisely,

$$(57) \qquad \frac{E(\max(S)) - E(|S|)}{E(\max(S))} \sim \frac{c}{\rho^a}$$

was found to hold for constants c and $a \approx 0.2$. Results confirming the conjecture[†] were also obtained for the case of two equally likely item sizes, 1 and $\sqrt{2}$.

7. Diffusion Models

Diffusion models are a natural approach to coping with the combinatorial complexities of the storage problems covered in earlier sections. In such models the earlier jump processes are approximated by continuous-path processes called diffusion processes. The principal limitation of this approach is that good approximations usually require the systems being modeled to be in heavy traffic, a notion that will be defined more carefully later. Specifically, the theoretical basis of diffusion models is obtained by establishing continuous-path diffusion processes as heavy traffic limits of suitably normalized versions of the corresponding jump processes.

Practically speaking, an application of the method begins with the selection of the differential equations defining an appropriate continuous-path stochastic process and its boundary conditions. In principle, this choice need be guided only by the quality of the approximation obtained thereby. On the other hand, it is clearly desirable to confine oneself, when possible, to processes that can be exhibited as appropriate limits of the processes being modeled. As discussed later, this latter

---

† [Note added in proofs]   Recently, a number of new asymptotic results have been obtained for variants of our model of fragmentation; see [CofL] for a brief survey.

condition appears to limit such models to classical diffusion processes. However, we mention an example of processes, not known to be expressible as heavy traffic limits, which provide demonstrably good approximations to queueing processes.

Choosing differential equations entails the setting of certain parameters, e.g. the drift and diffusion coefficient of the diffusion process, Brownian motion. We can present certain guidelines, but in fact no systematic procedure exists. For a given set of parameters calculations based on the solution of the equations can be made to obtain the desired performance measures. By comparison with known results for special cases a trial and error approach can be used to refine the approximation.

As implied by our opening statement, a major advantage of the method is that we can expect to exploit the state-space collapse which is a characteristic feature of continuous-path limits [Rei1]. As illustrated later, this occurs when the effect of much of the information in states of the initial process becomes negligible in the limit producing a diffusion process. As a second advantage, when exact methods are feasible only under approximate distributional assumptions, e.g. exponential interarrival or service times, a diffusion model in which such assumptions need not be made may yield results that are at least as good. Finally, measures of transient behavior, such as first-passage times, may be substantially easier to compute in the continuous-path model.

In the sequel we present the fundamental heavy traffic limit theorems developed in [CofR1] which are based on the storage systems defined in section 3. Following [CofR2] these results are supplemented by discussions of a number of related modeling issues: the circumstances under which the approximation can be expected to be useful for a given system; other, structurally more general systems to which the approximation applies; defining continuous-path approximations for stable systems; and finally, the calculation of performance measures.

The Mathematical Model. Along with notational matters we now introduce the processes fundamental to the models studied later. Initially, to fix ideas we choose a simple GI/G/1 queueing system: Items arrive according to a given but arbitrary renewal process $A(t)$ and join a FIFO queue. The mean and variance of the interarrival periods are denoted by $1/\lambda$ and σ_A^2, respectively. We let (B_i, R_i), $i \geq 1$, denote the size and the service time, respectively, of the ith arriving item.

The random variables $\{(B_i, R_i), \ i \geqslant 1\}$ are assumed to be i.i.d., and have in common the joint distribution function $G(x, y)$; the corresponding marginal distributions are denoted by $G_B(x)$ and $G_R(y)$. The mean and variance of the item size distribution are given by b and σ_B^2, respectively. The corresponding service-time parameters are $1/\mu$ and σ_R^2.

The service or departure process $S(t)$ denotes the number served in the interval $[0, t]$. The number of items in the system at time t is denoted by $Q(t) = A(t) - S(t)$. The total amount of storage occupied at time t is simply the sum of the item sizes and is denoted by $M(t)$. These latter two processes will be termed the queue-length and storage processes, respectively.

In the next section we establish a heavy-traffic limit theorem for a suitable normalization of $M(t)$. The theory of weak convergence [Bil] provides the framework for this result. The reader more interested in the engineering aspects of the diffusion approximation may choose to skip the proof of this theorem; however, it is important to bear in mind the mechanism by which convergence to a continuous-path limit is achieved.

Before proceeding a comment on applications is in order. As in the buffer systems of chapter 2, it is typical in the storage systems being modeled here that an arriving item has to be loaded into a storage device before it can be serviced. Unloading processes at service terminations may also be required. Since our mathematical model does not represent such processes explicitly, the model applies chiefly to those systems whose operation overlaps these processes with multiprogrammed job execution, which we find to be consistent with our heavy traffic assumptions, or to systems allowing us to include these delays as part of the service times, as in a delay (infinite server) system for example.

A Limit Theorem for the GI/G/1 Storage System. A continuous-path stochastic process having independent, normally distributed increments with mean ct and variance $\sigma^2 t$ is known as Brownian motion with drift[†] c and diffusion coefficient σ^2.

---

† Since c is common notation for Brownian motion drift, we use the upper case C later in this section to denote storage capacity.

We let $BM(c,\sigma^2)$ denote this process. (Except where noted we assume zero initial states for the diffusion processes of this section.) Imposing an instantaneously reflecting barrier at the origin defines the process known as reflected Brownian motion and is denoted $RBM(c,\sigma^2)$. The differential equations describing diffusion processes are deferred until needed for later calculations.

To formulate a continuous-path limit we proceed in the usual way to define a sequence of GI/G/1 systems indexed by $n \geqslant 1$, with $Q_n(t)$, $M_n(t)$ and $S_n(t)$ denoting the processes and $\lambda(n)$, $\sigma_A^2(n)$, etc. denoting the parameters in the nth system. We assume the following finite limits as $n \to \infty$

$$\lambda(n) \to \lambda \ , \ \sigma_A^2(n) \to \sigma_A^2$$

(58)
$$\mu(n) \to \mu \ , \ \sigma_R^2(n) \to \sigma_R^2$$

$$b(n) \to b \ , \ \sigma_B^2(n) \to \sigma_B^2$$

Also, for all $n \geqslant 1$ we assume that finite third moments exist for the interarrival and service time distributions of the nth system. It is possible to weaken this requirement slightly, but as given it is normally met in applications.

Although the index n does not have a direct physical interpretation, a discrete sequence of systems is used for two reasons. First, this formulation allows a unified treatment of the three cases $\lambda(n)-\mu(n) \to 0$ from above or from below, and $\lambda(n)-\mu(n) = 0$. The case where $\lambda(n)-\mu(n) \to 0$ from below is typically of most interest in applications, because it corresponds to a sequence of stable systems. An appropriate physical interpretation of the index n for this case is considered later.

The second reason for using the index n is that it brings out the relation of these results to classical central limit theorems.

Following the general theory we now introduce normalized jump processes, $\hat{Q}_n(t)$ and $\hat{M}_n(t)$, that reduce jump sizes by a factor of $\sqrt{n}$ but, by means of a time compression, increase the number of jumps in any given time interval by a factor of n. Specifically, over any given finite interval $[0,T]$ define

(59)
$$\hat{Q}_n(t) = \frac{Q_n(nt)}{\sqrt{n}}, \ \ \hat{M}_n(t) = \frac{M_n(nt)}{\sqrt{n}}, \ \ 0 \leqslant t \leqslant T .$$

The following limit theorem shows that if the difference in the arrival rate $\lambda(n)$

and the service rate $\mu(n)$ approaches zero at an appropriate rate as $n \to \infty$, then the (non-Markov) processes $\hat{M}_n(t)$ converge to a Markov, continuous-path limit.

In the sequel, weak convergence of stochastic processes is denoted as usual by the symbol $\to$. For economy of notation we also use this symbol to denote convergence in distribution of a sequence of real-valued random variables. Also, without explicit statement, convergence is always assumed to occur as $n \to \infty$.

Theorem 1. Suppose the parameters $\lambda(n)$ and $\mu(n)$, $n \geqslant 1$, are such that as $n \to \infty$

(60) $$(\lambda(n)-\mu(n))\sqrt{n} \to c \ .$$

Then

(61) $$\hat{M}_n \to b \ RBM(c,\sigma^2) = RBM(bc,b^2\sigma^2) \ ,$$

where

(62) $$\sigma^2 = \lambda^3\sigma_A^2+\mu^3\sigma_R^2 \ .$$

Proof.[†] According to our definitions

$$M_n(t) = \sum_{i-S_n(t)+1}^{A_n(t)} B_i(n) - \sum_{i-S_n(t)+1}^{S_n(t)+Q_n(t)} B_i(n)$$

for any sample sequence $\{(B_i(n), R_i(n)), i \geqslant 1\}$ in the nth system. Introducing the normalized storage process, we have

(63) $$\hat{M}_n(t) = \frac{1}{\sqrt{n}} \sum_{i-S_n(nt)+1}^{S_n(nt)+Q_n(nt)} B_i(n) \ .$$

Our next step shows that

(64) $$\sup_{0\leqslant t\leqslant T} \left| \frac{1}{\sqrt{n}} \sum_{i-S_n(nt)+1}^{S_n(nt)+Q_n(nt)} B_i(n) - b(n)\hat{Q}_n(t) \right| \to 0 \ .$$

---

† A useful, brief survey of the weak-convergence results that we need can be found in [Rei1].

A direct application of the converging-together theorem then establishes that if $\hat{Q}_n \rightarrow \hat{Q}$ then $\hat{M}_n \rightarrow b\hat{Q}$.

To prove (64) we use two additional results from weak-convergence theory.

(i)   Let

(65) $\beta_n(t) = \dfrac{1}{\sqrt{n}} \sum\limits_{i=1}^{\lfloor nt \rfloor} (B_i(n) - b(n)), \quad n \geq 1, 0 \leq t \leq T .$

By our assumptions on $\{B_i(n), i \geq 1\}$ we have that $\beta_n \rightarrow \beta = BM(0, \sigma_B^2)$. (This is essentially Prohorov's extended central limit theorem [Pro].)

(ii)   If for some process $\psi_n(t)$, $\sup\limits_{0 \leq t \leq T} \psi_n(t) \rightarrow 0$, and if $\beta_n \rightarrow \beta = BM$, then as a special case of the random-time-change theorem we have

(66) $\sup\limits_{0 \leq t \leq T} \left| \beta_n(t + \psi_n(t)) - \beta_n(t) \right| \rightarrow 0 .$

Now expanding (63) we can write

(67) $\dfrac{1}{\sqrt{n}} \sum\limits_{i = S_n(nt)+1}^{S_n(nt)+Q_n(nt)} B_i(n) =$

$\dfrac{1}{\sqrt{n}} \sum\limits_{i = S_n(nt)+1}^{S_n(nt)+Q_n(nt)} (B_i(n) - b(n)) + b(n)\hat{Q}_n(t), \, 0 \leq t \leq T .$

Let

$$\phi_n(t) = \frac{S_n(nt)}{n} - \mu t .$$

Then by (65)

$\dfrac{1}{\sqrt{n}} \sum\limits_{i = S_n(nt)+1}^{S_n(nt)+Q_n(nt)} (B_i(n) - b(n)) = \beta_n(\mu t + \phi_n(t) + \dfrac{Q_n(nt)}{n}) - \beta_n(\mu t + \phi_n(t))$

$= \beta_n(\mu t + \phi_n(t) + \dfrac{Q_n(nt)}{n}) - \beta_n(\mu t)$

$- [\beta_n(\mu t + \phi_n(t)) - \beta_n(\mu t)] .$

Clearly, $\sup\limits_{0 \leq t \leq T} Q_n(nt)/n = \sup\limits_{0 \leq t \leq T} \hat{Q}_n(t)/\sqrt{n} \rightarrow 0$. Moreover, $\sup\limits_{0 \leq t \leq T} |\phi_n(t)| \rightarrow 0$

by the weak law of large numbers. Thus, identifying $\psi_n(t)$ in (66) with $\phi_n(t) + Q_n(nt)/n$, and then accounting for the change in time scale, we can apply (66) to obtain

$$(68) \qquad \sup_{0 \leqslant t \leqslant T} \left| \frac{1}{\sqrt{n}} \sum_{i=S_n(nt)+1}^{S_n(nt)+Q_n(nt)} (B_i(n) - b(n)) \right| \to 0 .$$

Use of this result in (67) implies (64) directly. Applying the classical diffusion approximation for the GI/G/1 queue [IglW], viz. $\hat{Q}_n \to \hat{Q} \to RBM(c, \sigma^2)$ (where c and σ^2 are given by (3) and (5)), we thus obtain $\hat{M}_n \to b\hat{Q} = RBM(bc, b^2\sigma^2)$ ∎.

Note that (63) is equal in distribution to

$$\frac{1}{\sqrt{n}} \sum_{i=1}^{Q_n(nt)} B_i(n) = \frac{1}{\sqrt{n}} \sum_{i=1}^{\sqrt{n}\hat{Q}_n(t)} B_i(n) ,$$

where $\{B_i(n), i \geqslant 1\}$ is a sequence of i.i.d. random variables with the distribution function $G_B^{(n)}(x)$. Thus, the essence of the theorem is the extended weak law of large numbers: If $\hat{Q}_n \to RBM$, then as $n \to \infty$

$$\frac{\sum_{i=1}^{\sqrt{n}\hat{Q}_n} (B_i - b)}{\sqrt{n}} \to 0 .$$

Also, note that the variance of item sizes does not enter into the continuous-path limit; only the expected value, b, appears in (61). The existence of this moment was required in our use of the limit $\beta_n \to BM(0, \sigma_B^2)$; however, it is an open question whether the existence of σ_B^2 is necessary in a proof of $\hat{M}_n \to b\hat{Q}_n$.

Let us now consider the nature of the approximation. It follows simply from our approximation of a jump process $M_n(t)$ by a continuous-path process, RBM, that the quality of the approximation improves as we consider systems in which the times between jumps and the sizes of the jumps themselves are rather small. Thus, large arrival and departure rates, λ and μ, are implied along with relatively small expected block sizes, b. Also, in order that the continuous limit not degenerate, we required that the difference in arrival and service rates approach 0 at a rate that produces a diffusion with finite drift. This obviously suggests that the system being modeled also be operating in heavy traffic, where $\rho = \lambda/\mu$ is close to 1.

Under the above circumstances the limit theorem justifies the approximation of the distribution of the normalized storage process by the distribution of an appropriate RBM. However, the limit theorem does not yield information on how close ρ must be to 1 to achieve a given degree of accuracy in the approximation. This information must come from results on rates of convergence.

In modeling a specific stable system by an appropriate limit diffusion process one must determine the index, n, to be associated with the given system. Since the index n is not directly related to physical properties of the system, it is convenient to express the passage to the limit, $n \to \infty$, in terms of a limit $\rho \to 1$. The choice $n = (1-\rho)^{-2}$ is a convenient one. Equation (60) then leads to $c = -\mu$ as a natural choice of drift for our approximating diffusion. The choice of variance is not uniquely determined by the limit theorem, since multiplying the variance term by ρ or $1/\rho$ yields the same limit as $n \to \infty$. The most reasonable approach is simply to use the variance term that appears in the limit theorem. After making these choices one simply inverts the normalization to obtain the approximation. In particular, from (59) one obtains

(69)
$$M(t) \stackrel{d}{\approx} \frac{\hat{M}(t(1-\rho)^2)}{1-\rho} \ .$$

Extensions to Other Systems. By applying diffusion limits for other queueing processes, we can exploit $\hat{M}_n \to b\hat{Q}$ in Theorem 1 and obtain similar limits for storage processes. For example, suppose we impose a bound (waiting-room constraint), $N(n)$, on the number of items that can be in the nth system at any given time; arrivals when there are $N(n)$ in the system are simply lost. We use primes in our notation for processes in this system. It is well-known that if $N(n)/\sqrt{n} \to N$ then $\hat{Q}'_n(t) \to RBM(c,\sigma^2,N)$, where c and σ^2 are defined as in Theorem 1, and where $RBM(c,\sigma^2,N)$ denotes Brownian motion reflected at both the origin and at N. This result along with Theorem 1 proves

Theorem 2

Under the hypotheses of Theorem 1 and the assumption $N(n)/\sqrt{n} \to N$, we have

$$\hat{M}'_n \to RBM(bc,b^2\sigma^2,bN) \ .$$

Another system of obvious importance is the GI/G/1 queue with a finite storage capacity, say C. In this system an arrival at time t is admitted if and only if its size does not exceed $C-M(t)$; otherwise, the arriving job is simply lost. Intuitively, one again expects an appropriate diffusion approximation to be Brownian motion with reflecting barriers at 0 and C. Indeed, with the same normalizations and assumptions as before one expects $\hat{M}_n'' \rightarrow RBM(bc, b^2\sigma^2, C)$, where the double prime denotes the storage limited system. However, this limit has yet to be proved.

Effectively, this would imply that the sequence of processes $\{\hat{M}_n'(t)\}$ for systems with queue-length limits $\sqrt{n}C/b$ converges to the same diffusion as the sequence $\{\hat{M}_n''(t)\}$ for systems with storage capacities C. A proof of this appears to be complicated necessarily by the behavior of $M_n''(t)$ near the barrier C. In the limit $n \rightarrow \infty$ this barrier becomes a reflecting barrier. However, if we assume a continuous item size distribution, the process' reaching this barrier in any finite time interval is an event with probability 0 for any finite n. Thus, a comparison of $M_n''(t)$ with $M_n'(t)$, for which the diffusion limit is known, seems to face non-trivial combinatorial questions.

Finally, we note that the class of service disciplines for which GI/G/1 diffusion limits are known extends substantially beyond the simple FCFS rule. In particular, a similar diffusion limit exists for the so-called *multi-class feedback queue* [Rei2]. To define these systems for our purposes here, let J be the number of item classes. Items in the J classes arrive according to independent renewal processes $A_j(t)$; i.e. $A(t)$ is the superposition of the $A_j(t)$, $1 \leqslant j \leqslant J$. The ith arriving item of class j is characterized by a size $B^{(i)}(j)$ and a service requirement given by the sequence of random variables $R_{i1}(j)$, $R_{i2}(j),...,R_{i,T_i(j)}(j)$, where $T_i(j)$ is also a random variable. Such an item must "visit" the server $T_i(j)$ times, receiving on the kth visit $R_{ik}(j)$ uninterrupted units of service time. The service discipline is first-come-first-served, one visit at a time; i.e. after each visit, except the last one, items are fed back to the end of the queue to await their next visit. Note that single-class FCFS and processor sharing are easily shown to be special cases of this system.

Performance Measures. Performance evaluation using a given diffusion approximation reduces essentially to the study of certain differential equations. A common point of departure is the (forward) diffusion equation

(70)
$$\frac{\partial p}{\partial t} = -c \frac{\partial p}{\partial x} + \frac{\sigma^2}{2} \frac{\partial^2 p}{\partial x^2} ,$$

describing the displacement density for Brownian motion with drift c and diffusion coefficient σ^2; i.e. with our assumed initial condition $p(x,0) = \delta(x)$, $p(x,t)$ is the density function for the (storage) displacement from 0 in time t. If only the initial condition is specified, then $p(x,t)$ is simply the normal density with mean ct and variance $\sigma^2 t$ associated with unrestricted Brownian motion.

For reflected Brownian motion, (70) must be solved under the requirement that

(71)
$$\frac{\sigma^2}{2} \frac{\partial p}{\partial x} - cp = 0$$

at the boundaries. Solutions for these cases take rather complicated forms and are omitted here (e.g. see [SweH] for these expressions along with computational details). In the case of two reflecting barriers, at 0 and N say, a stationary density exists and is given by

(72)
$$p(x) = -\frac{2c}{\sigma^2} \frac{e^{2cx/\sigma^2}}{1 - e^{2cN/\sigma^2}} , 0 \leqslant x \leqslant N .$$

For a single reflecting barrier at the origin a stationary density exists only if the drift c is negative (i.e. towards the origin). For this case it suffices to take the limit $N \to \infty$ in (72) to obtain the exponential density

(73)
$$p(x) = -\frac{2c}{\sigma^2} e^{2cx/\sigma^2} , x \geqslant 0 .$$

Note that these results are obtainable from the ordinary differential equations resulting from (70) and (71) when the functional dependence on time is eliminated. Moments of the stationary storage occupancy distribution are easily calculated in the usual way.

First passage times to an upper boundary are obviously important performance measures of the system. Although density functions are again rather complicated (see [CoxM]), moments can be obtained rather easily as follows. For the process $X(t) = RBM(c,\sigma^2)$ define the first passage time

$$\tau(x,N) = \inf\{t \geqslant 0 | X(t) \geqslant N, X(0) = x\} ,$$

let $m_i(x,N)$ denote the ith moment, $E[\tau^i(x,N)]$, and define $m_0(x,N) = 1$. Then it is not difficult to show that

(74)
$$\frac{\sigma^2}{2} \frac{d^2 m_i}{dx^2} + c \frac{dm_i}{dx} = -im_{i-1} , \quad i \geqslant 1 ,$$

with the boundary conditions $m_i(N,N) = 0$ and $\dfrac{dm_i}{dx} = 0$ at $x = 0$. Solutions for the first two moments can be obtained routinely; e.g. for an initial state $x = 0$ we obtain

(75)
$$E[\tau(0,N)] = \frac{N}{c} - \frac{\sigma^2}{2c^2}[1-\exp(-2cN/\sigma^2)] , \quad c \neq 0$$
$$= \frac{N^2}{\sigma^2} , \quad c = 0 .$$

We conclude this analysis of GI/G/1 systems with the observation that tractable continuous-path processes have been proposed which reside at the origin, whenever it is reached, for a time interval governed by some given distribution (the exponential distribution is normally chosen for its Markov property). At the end of these holding times the process makes an instantaneous jump back into the interior. The magnitude of this jump can again be a suitably chosen random variable. These return processes were originated by Feller [Fel2] and appear, at least superficially, to be natural models of stable storage systems; the holding times are interpreted as idle periods and determined by residual interarrival periods. Indeed, remarkably good agreement has been found between results for queueing processes obtained by simulation and those obtained from analyses of return processes [Gel2] (see [GelM] and [AveGK] for numerical results). Extensions to storage processes have not yet been studied in any detail.

From a mathematical standpoint, *RBM* can be exhibited straightforwardly as an appropriate heavy-traffic limit of return processes. However, in their intended use, return processes suffer from the fact that it appears impossible to establish them as limits of the systems (stable queues) being modeled. In particular, the heavy-traffic assumption appears to be inherent in any continuous-path limit. In other words, to be consistent with a continuous-path limit, we need the interarrival and service times (and hence absorption times at a boundary) to vanish in the limit.

Delay or Infinite-Server Storage Systems. In this system all items immediately commence their service times at their times of arrival. Thus, the departure rate is proportional to the number in system, and the system is stable for all $0 < \lambda < \infty$ and $0 < \mu < \infty$. Exact results are not hard to find for the Markov (i.e. $M/M/\infty$) system with unbounded storage. For this purpose, and for the remainder of this section, it is convenient to assume that possible block sizes comprise the discrete set $\{b_1, b_2, ..., b_r\}$, where r need not be finite. Note that this assumption incurs no loss in generality for practical systems. However, r may be prohibitively large for computational purposes, so that a coarser discretization may have to be adopted as an approximation.

Let $G_B(x)$ be defined by the probability mass function $\{\lambda_i/\lambda, 1 \leqslant i \leqslant r\}$, where λ_i can be interpreted as the parameter of the Poisson arrival process for items with sizes, b_i. Let μ_i be the service rate of items having size b_i.

Define the process $Q^{(i)}(t)$ as the number of items of size b_i, $1 \leqslant i \leqslant r$, in the system at time t, and let $M^{(i)}(t) = b_i Q^{(i)}(t)$ be the storage they occupy. Clearly, the processes $\{Q^{(i)}(t), 1 \leqslant i \leqslant r\}$, and hence the processes $\{M^{(i)}(t), 1 \leqslant i \leqslant r\}$ are mutually independent. As noted previously the classical result for the stationary distribution of $Q^{(i)}(t)$ is given by

$$Q^{(i)}(t) \;\rightarrow\; P_i \,,$$

as $t \rightarrow \infty$, where P_i denotes a random variable governed by the Poisson frequency function with parameter λ_i/μ_i. Thus, as $t \rightarrow \infty$

$$M^{(i)}(t) \;\rightarrow\; b_i P_i \,,$$

and hence

$$M(t) = \sum_{i=1}^{r} M^{(i)}(t) \;\rightarrow\; \sum_{i=1}^{r} b_i P_i \,,$$

where the P_i, $1 \leqslant i \leqslant r$, are mutually independent.

We turn now to the Ornstein-Uhlenbeck process as a model of certain infinite-server queues in heavy traffic. One of the early uses of this diffusion process was as a model of particle motion when a restoring force acting on the particle increased proportionally with its displacement from a given central position. Since the tendency of $Q(t)$ towards the "balance point", λ/μ, in an infinite-server system increases proportionally with the distance from this point, it is natural to consider the Ornstein-Uhlenbeck process as a diffusion approximation for a suitable normalization of $Q(t)$ under heavy traffic assumptions.

Assuming that the initial state is 0, the normal displacement density, $p(x,t)$, of this process satisfies

$$(76) \qquad \frac{\partial p}{\partial t} = -cx \, \frac{\partial p}{\partial x} + \frac{\sigma^2}{2} \, \frac{\partial^2 p}{\partial x^2} \,,$$

$$p(x,0) = \delta(x) \,,$$

and has mean 0 and variance $(\sigma^2/2)(1-e^{-2ct})$, where cx and σ^2 are the infinitesimal mean and variance in analogy with the diffusion equation (70). Since c is non-negative, the stationary density exists and has mean 0 and variance $\sigma^2/2$. For the special case $c = 0$, the process degenerates to unrestricted Brownian motion with zero drift. Calculations of first-passage times can be found in [KeiR]. Whitt [Whi] gives a full account of the applications of this process in the approximation of infinite-server queues under heavy traffic. In the sequel we let $OU(c,\sigma^2)$ denote the Ornstein-Uhlenbeck process defined by (76).

Under the assumption of exponential service times, limit theorems of Iglehart [Igl1] and Borovkov [Bor] lead to continuous-path approximations of corresponding storage processes. In particular, using our earlier notation we can define a sequence of GI/M/∞ storage systems with $\lambda(n) = n\lambda$ and $\mu(n) = \mu$, in the nth system, $n \geqslant 1$. For the normalized process

$$(77) \qquad \hat{Q}_n(t) = [Q_n(t)-n\rho] \, / \, \sqrt{n} \,,$$

where $\rho = \lambda/\mu$, we have

$$(78) \qquad \hat{Q}_n \rightarrow OU(\mu,\lambda+\lambda^3\sigma_A^2) \,,$$

where it is assumed that $\hat{Q}_n(0) \rightarrow \hat{Q}(0)$, and $\hat{Q}(0)$ is a proper random variable. Note that the infinitesimal variance reduces to 2λ in the special case of Poisson arrivals ($\sigma_A^2 = 1/\lambda^2$).

Exploiting our discretization of item sizes, the extension to storage systems is but a short step from (78). We assume that each distinct item size has its own renewal input process with rate λ_i and interarrival-time variance σ_i^2. This superposition arrival process gives rise to a system denoted by the symbol $\Sigma GI/M/\infty$.

Theorem 3

Using our earlier notation and the normalization $\hat{M}_n^{(i)}(t) = b_i \hat{Q}_n^{(i)}(t)$, we have for the $\Sigma GI/M/\infty$ storage process $\hat{M}_n(t) = \Sigma_{1 \leqslant i \leqslant r} \hat{M}_n^{(i)}(t)$,

$$\hat{M}_n \rightarrow \sum_{i=1}^{r} b_i OU_i ,$$

where $OU_i = OU(\mu_i, [\lambda_i + \lambda_i^3 \sigma_i^2])$.

Remarks

1. We could also consider the system in which all arrivals are contained in a single renewal process, so that the arrival process for each distinct block size is a thinning of the original renewal process. This introduces a dependence between the component OU processes, except for Poisson arrivals when the two models are distributionally equivalent.

2. $\hat{M}_n$ converges to a linear combination of diffusion (i.e. Markov) processes, and therefore to a Gaussian process, but the limit is not itself a Markov process. The vector process $<OU_1,...,OU_r>$ is, however, a multivariate diffusion process.

3. Observe that the continuous-path limit now depends on the entire distribution of block sizes and not just on the mean value as in Theorems 1 and 2.

From results of Borovkov for the $GI/G/\infty$ system [Bor] we can develop a corresponding continuous-path limit

$$\hat{M}_n \rightarrow \sum_{i=1}^{r} b_i X_i ,$$

where the $X_i(t)$ (and hence the sum $\Sigma_{1 \leqslant i \leqslant r} b_i X_i(t)$) are Gaussian processes. However, the X_i will be Markov, viz. OU, processes only if the service times are exponential (the $GI/M/\infty$ systems of Theorem 3).

The above results are in fact special cases of a more general result due to Iglehart [Igl2]. Iglehart's result can be applied directly to the storage process with general (not necessarily discrete) size distributions, yielding the form of the Gaussian limit process. His results also apply to the system described in Remark 1. The above model has been generalized to systems in which the intensity of the arrival process depends on the number in system. The analysis uses a martingale approach [KogL].

The loss of the Markov property for general service times originates with the normalization needed for a continuous-path limit. Specifically, according to (77) service times do not converge to 0 as they do according to (59) in the GI/G/1 case. Thus, the lack of a Markov property in the M/G/∞ queue persists in the limit. With the goal of simplifying computations in applications, Whitt [Whi] has specialized the general limit theorems to service distributions of phase type, which can be used to approximate any given distribution arbitrarily closely by sums of exponentials. Thus, each Gaussian limit process can be expressed as a multivariate Markov process.

We conclude by mentioning a few of the numerous open problems in the study of continuous-path approximations to storage processes in heavy traffic. The conjectured limit concerning storage limited GI/G/1 systems is one such problem. Also, effective uses of the Gaussian processes just introduced have yet to be explored in detail. The calculation of first-passage times, for example, appears to be difficult, owing to the absence of the Markov property. Finally, there is the problem of modeling storage-limited processes in infinite-server systems by continuous-path processes with reflecting barriers. In this connection, we note that the normalization in (77) precludes a continuous-path limit restricted to a finite interval in the space dimension.

Chapter 4

Paged Two-Level Memories:
Theory and Analysis of Algorithms

1. General Definitions

We consider the problems that arise in automating the exchange of information between the two principal levels of computer memory. As in the introduction of section 1.3, these levels are referred to as the main memory, and the (slower) auxiliary or secondary memory. The term memory by itself refers to main memory. To organize the exchange, the information is subdivided into blocks of variable length, called segments, or blocks of fixed length, called pages. Only entire segments or pages can be transferred from one memory level to the other.

A *replacement algorithm* defines the set of segments or pages in the faster main memory. In particular, it determines which pages or segments are to be replaced, when necessary, in order to make room for a page or segment that has been referenced but not found in main memory. Clearly, the frequency of the more time consuming references to auxiliary memory, and consequently the processing rate in the two-level memory as a whole, depend heavily on the replacement algorithm chosen.

In what follows we concentrate on a widely used mechanism for storage allocation which is known as virtual memory with paging. Specifically, we consider a mathematical model of page replacement control in a two-level memory in which the basic unit of stored and transmitted information is a page.

With a fixed amount of memory assigned to a program, the minimum number of references to the auxiliary memory is achieved by an algorithm that always replaces that page in main memory which is not to be referenced for the longest period of time. However, such an algorithm is unrealizable, since advance information is normally not available on future page references. Thus, the main problem that arises in the synthesis of replacement algorithms is the identification of algorithms that will give an acceptably low frequency of references to auxiliary memory for the greatest variety of programs with structures not known in advance.

94

Under multiprogramming conditions the amount of main memory occupied by programs is either fixed or is defined by the dynamics of the running programs. Replacement algorithms designed for operating under a fixed partition of memory are illustrated below.

1. Random Replacement (RR): random selection is used for determining the page to be replaced; i.e. each page is equally likely to be the one replaced.

2. First-In-First-Out (FIFO): that page is replaced which has been in memory for the longest time.

3. Least-Recently-Used (LRU): that page is replaced whose last reference (usage) was earliest among those pages in memory.

Of the algorithms which allow a variable amount of memory to be occupied by the program, the Working Set (WS) algorithm is the best known. With this algorithm, at any given time t only those pages of the program will be retained in memory which have been referenced during the time interval $(t-\tau, t)$, where τ is a parameter. In contrast to such algorithms as RR, FIFO, and LRU, the WS algorithm can increase or decrease the memory size occupied by the program, depending on program "behavior". Therefore, the WS algorithm can be used for dynamic allocation of memory to several programs.

Consider a given set of program pages $X = \{1, 2,..., n\}$. Assume that all n pages are permanently in auxiliary memory and that $m < n$ pages may be stored in main memory. We describe the page referencing activity during execution of a program by the reference string $\mathbf{x} = x_1, x_2,...$. If $x_t = i$, we understand that the program references page i at time t. If on being referenced, a page is absent from main memory, we have what is referred to as a *page fault*. Let $X_t \subseteq X$ be the set of pages of the program in main memory at time t. We use a superscript, A for example, to denote that the replacement rule generating X_t^A is algorithm A.

Definition 1. Replacement algorithm A is said to be *physically realizable* if at each time instant $t = 1, 2,...$ it determines the set X_t^A as a function (possibly randomized) only of the observed history x_i, $0 \leqslant i \leqslant t-1$, and x_t.

The algorithms RR, FIFO, LRU, and WS are examples of physically realizable replacement algorithms. The next definition introduces an important class of replacement algorithms.

Definition 2. Let the memory assigned to a program consist of m pages. With a fixed amount of memory assigned to a program, a replacement algorithm A is a *demand replacement algorithm* if it determines the set X_t^A in the nontrivial case $|X_0^A| = m$ in accordance with the formula

$$
X_t^A = \begin{cases} X_{t-1}^A & \text{if } x_t \in X_{t-1}^A , \\ X_{t-1}^A \backslash y_t^A + x_t & \text{if } x_t \notin X_{t-1}^A , \end{cases}
$$

where y_t^A is the page replaced by algorithm A.

RR, FIFO, and LRU are examples of demand replacement algorithms.

In implementing the FIFO and LRU rules the set of pages, X_t, retained in main memory at any time t is represented by an ordered list. This list is called the memory state. For instance, if $X_{t-1} = (i_1, i_2, ..., i_m)$ is the memory state at time $t-1$, then for FIFO, page i_k entered the memory earlier than i_{k-1} for each k, $1 < k \leqslant m$. With the FIFO rule the memory state changes only at times of page faults. If at time t a requested page is $x_t \notin X_{t-1}$, then page i_m is replaced and the new memory state is $X_t = (x_t, i_1, i_2, ..., i_{m-1})$. In the case of the LRU algorithm the memory state may also change when reference is made to pages already present in memory. If $X_{t-1} = (i_1, i_2, ..., i_k, ..., i_m)$ and $x_t = i_k$, then $X_t = (i_k, i_1, ..., i_{k-1}, i_{k+1}, ..., i_m)$. If a page fault occurs, then the memory state for LRU changes in accordance with the same rule as for FIFO. We conclude with a definition of a class of Markov replacement algorithms with finite memory, a subset of physically realizable demand replacement algorithms.

Definition 3. We will call replacement algorithm A *Markovian with finite memory* [AveK3] if $X_t^A = f(X_{t-1}^A, x_t)$ where X_t^A is the memory state at time t under algorithm A. The rule $f(\cdot, \cdot)$ may or may not have randomization.

Convenient implementation is the advantage of such algorithms. Algorithms RR, FIFO, and LRU are examples of Markovian replacement algorithms with finite memory. The rule $f(\cdot, \cdot)$ used for FIFO or LRU is nonrandomized (i.e. deterministic), and for RR it is randomized. The LFU algorithm, which always replaces the least frequently used page, is an example of an algorithm without finite memory. Indeed, to implement LFU, we have to maintain and store the reference frequencies for all pages in both main and auxiliary memory. It is easy to see that X_t^{LFU} cannot be determined by X_{t-1}^{LFU} and x_t alone.

2. A Class of Self-Organizing Replacement Algorithms

Experiments have shown that most programs have a relatively small subset of pages with the property that, if these pages are kept in main memory, the frequency of page faults remains relatively low. Such a subset is often referred to as the working set of the program. Here, we shall use the term *locality set* in order to distinguish it from the working set introduced in the previous section. We shall not define a locality set formally. However, for the informal discussion below this is not necessary.

The locality set may vary during program execution, and it is therefore desirable for a replacement algorithm to have the following two properties, which are in part contradictory: The replacement algorithm should retain in main memory the pages of the locality set, and it should rapidly update memory when the locality set is changed.

The simplest realizable algorithm is the FIFO rule, which is effective only in that part of memory being rapidly updated; it does not distinguish in X_t the pages of the locality set that are referenced more frequently than other pages. In contrast to FIFO the LRU algorithm updates X_t on referencing the pages in memory, and under certain conditions will retain the locality-set pages in memory. For example if all locality set pages are referenced in the time interval between every two consecutive references to auxiliary memory, then the locality set is always contained in X_t^{LRU}. However, LRU is similar to FIFO in that sequences of single references to different pages in auxiliary memory may erase all locality-set pages from memory. Since the page just introduced into memory takes the first position in the list $X_t^{\text{LRU}}(X_t^{\text{FIFO}})$ a few single references to pages not in main memory may cause erasure of an equal number of locality-set pages. The WS algorithm has a similar drawback. These disadvantages motivate the following class of algorithms.

The self-organizing algorithms defined below depend on a finite number of parameters; adaptive selection of these parameters permits a combination of fast updating for a part of memory with retention in memory of those pages most frequently required (i.e. those contained in the locality set). This provides an additional mechanism for reducing the frequency of page faults.

Let $X_t = (i_1, i_2, ..., i_{m_t})$ be an ordered list of pages that defines the memory state at time t, where m_t is the memory size assigned to the program at time t. To define the replacement algorithm it is necessary to state the rules for transforming X_t to X_{t+1}. In general, a self-organizing replacement algorithm determines the position of page x_{t+1} in X_{t+1} with the aid of h disjoint priority subsets (lists) $L_1, L_2, ..., L_h$, whose union is X_{t+1}. When a page x_{t+1} is introduced into memory it occupies the first (highest priority) position on list L_h; if x_{t+1} is already in memory and occupies a position in subset $D_i \subseteq L_i$, $i \geqslant 2$, then this page is assigned the first position in the list with the next higher priority, L_{i-1}. The subsets D_i provide for additional flexibility. If D_i is a proper subset of L_i, then pages that belong to $L_i \backslash D_i$ do not leave the priority set L_i when they are referenced. Thus an appropriate choice of D_i creates a reduction in time during which temporarily (or rarely) used pages stay in memory.

In order to include in the class of self organizing replacement algorithms those of WS type, we introduce two sets of parameters. In the first set the parameters T_i, $i = 1, 2, ..., h$, specify the maximum permissible storage time[†] of a page in memory without being referenced under the condition that its position is in list L_i, $i = 1, 2, ..., h$. In the second set are the parameters, μ_i, that determine the maximum number of elements in the sets L_i, $i = 1, 2, ..., h$.

Definition 4. Assume that the rules are given for constructing the subsets $D_i \subseteq L_i$, $i = 1, 2, ..., h$, and let $X_t = (L_1, ..., L_h)$ be the memory state at time t. At time $t+1$ when page x_{t+1} is referenced, the new memory state under algorithm A is determined by means of the following steps [Kog3, Kog4].

Step 1. From each set L_r, $1 \leqslant r \leqslant h$, those pages not referenced for a time longer than T_r (i.e. not referenced in $[t-T_r+1, t+1]$) are removed.

---

† Time is measured only during the execution of the program.

Step 2. The remaining elements in the sets L_r are renumbered in such a way that the memory state X'_t after step 1 is $X'_t = (L'_1,..., L'_h)$, where the L'_r, $r = 1, 2,..., h$, are reduced to the form $L'_r = (i_{r1}, i_{r2},..., i_{rm_r})$; some of the subsets L'_r may be empty.

Step 3. The new memory state X_{t+1} differs from X'_t only by

Case 1: L_h if $x_{t+1} \notin X_t$;

Case 2: L_r, $1 \leqslant r \leqslant h$, if $x_{t+1} = i_{rs}$, $s = 1, 2, ..., m_r$, and $i_{rs} \in L_r \backslash D_r$, with $D_1 = \emptyset$,

Case 3: L_{r-1} and L_r, $2 \leqslant r \leqslant h$, if $x_{t+1} = i_{rs}$, and $i_{rs} \in D_r$.

Denote by τ_r the time since the last reference to page $i_{rm_r} \in L'_r$ and specify h integers $\ell_1, \ell_2, ..., \ell_h$ and h numbers $\alpha_1, \alpha_2, ..., \alpha_h$ where $0 < \alpha_i \leqslant 1$, $i = 1, 2,..., h$. (The meaning of these is explained later.) Then for case 1, L'_h is transformed into

$$L''_h = \begin{cases} (x_{t+1}, i_{h1}, ..., i_{hm_h}) & \text{if } \tau_h \leqslant \alpha_h T_h \text{ and } m_h < \mu_h , \\ (x_{t+1}, i_{h1}, ..., i_{hm_h-1}) & \text{if } \tau_h > \alpha_h T_h \quad \text{or} \quad m_h = \mu_h . \end{cases}$$

For case 2, L'_r is transformed into

$$L''_r = \begin{cases} L'_r & \text{if } 1 \leqslant s \leqslant \ell_r , \\ (i_{rs}, i_{r1}, ..., i_{r,s-1}, i_{r,s+1}, ..., i_{rm_r}) & \text{if } \ell_r < s \leqslant m_r . \end{cases}$$

For case 3, L'_{r-1} is transformed into

$$L''_{r-1} = \begin{cases} (i_{rs}, i_{r-1, 1},...,i_{r-1, m_{r-1}}) & \text{if } \tau_{r-1} \leqslant \alpha_{r-1} T_{r-1} \text{ and } m_{r-1} < \mu_{r-1} , \\ (i_{rs}, i_{r-1, 1},..., i_{r-1, m_{r-1}-1}) & \text{if } \tau_{r-1} > \alpha_{r-1} T_{r-1} \quad \text{or} \quad m_{r-1} = \mu_{r-1} , \end{cases}$$

and L'_r is transformed into

$$L''_r = \begin{cases} (i_{r1},...,i_{r,s-1},i_{r,s+1},...,i_{rm_r}) & \text{if } \tau_{r-1} \leqslant \alpha_{r-1} T_{r-1} \text{ and } m_{r-1} < \mu_{r-1} , \\ (i_{r-1,m_{r-1}},i_{r1},...,i_{r,s-1},i_{r,s+1},...,i_{rm_r}) & \text{if } \tau_{r-1} > \alpha_{r-1} T_{r-1} \quad \text{or} \quad m_{r-1} = \mu_{r-1} . \end{cases}$$

There are a number of self organizing replacement algorithms of practical and theoretical interest. For $h = 1$ and $\mu_h = \infty$, we obtain the modified WS algorithm [Smil], which for $\alpha_h = 1$ reduces to the WS algorithm. The parameter α_h enables one to bound the size of memory occupied by the program. The smaller α_h is, the smaller is the number of memory pages that the program occupies in the poor-performance case, when each of the pages is referenced once only. In a similar manner, by the choice of the parameter α_i, it is possible in general to control the size of the lists L_i, $i = 1, 2,..., h$, by not allowing them to increase excessively.

Setting $T_1 = T_2 = ... = T_h = \infty$ and $D_i = L_i$, $i = 2,3,...,h$, we obtain the class of replacement algorithms A_ℓ^h, introduced in [AveBK1, AveBK2], and defined by the $h+1$ parameters: $\mu_1, \mu_2, ..., \mu_h$ and $\ell = \ell_1$, i.e. $A_\ell^h = A_\ell^h(\mu_1, \mu_2, ..., \mu_h)$. For $h = 1$ and $\ell = m = \sum_{i=1}^{h} \mu_i$ we have the FIFO algorithm and for $h = 1$ and $\ell = 1$ we have the LRU algorithm. The algorithms A_ℓ^1, $1 < \ell < m$, are intermediate between FIFO and LRU and resemble the well-known algorithm "first in, if not used, then first out". If the positions of list X_t are numbered, say from left to right, then with the algorithm A_ℓ^1 the page in list position $i \leqslant \ell$ is changed as for FIFO, while that in position $i > \ell$ is changed as for LRU. The parameters ℓ_r in the definition of self organizing algorithms are interpreted in a similar manner within each of the subsets $L_r \backslash D_r$, $r = 1, 2, ..., h$.

Let us consider the case $h = m$, for which the parameter ℓ can take the value 1 only. According to algorithm A_1^m, a page just introduced into memory corresponds to position m in list X_t. If up to the next page fault no reference is made to this page, then it is replaced. If reference is made to a page in memory, its position i, $1 < i \leqslant m$, changes to $i-1$, and the page previously in position $i-1$ is placed in position i. Shifting from position m to position 1 can be likened to climbing a staircase of m steps, hence the term CLIMB algorithm.

For $1 < h < m$ the positions of list X_t split into h priority sets $L_1, L_2, ..., L_h$. The page just introduced into memory has the first position in the lowest-priority set L_h. After each new reference to this page its position successively shifts to the first position in sets of increasing priority. As an example, consider the replacement algorithm $A_2^3(3, 2, 3)$. The pages referenced are denoted by capital letters. Let $X_0 = (A, B, C, D, E, F, G, H)$ and let the reference string consist of the 20 elements

$$\mathbf{x} = (E, A, E, B, A, C, H, F, I, E, E, F, E, J, J, J, B, I, E, D) .$$

The operation of the replacement algorithm $A_2^3(3,2,3)$ is illustrated by the memory-state table in Fig. 1. Here, each column represents the memory state X_t at some time t. Arrows indicate the pages that have been replaced.

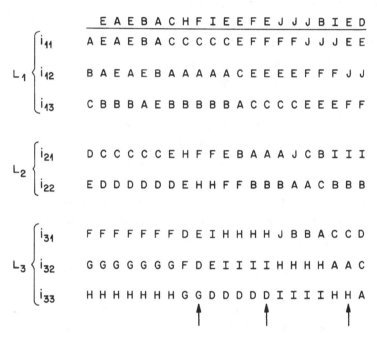

Figure 1 -   Example memory state table for A_2^3 (3,2,3).

The class of self-organizing algorithms widens appreciably the possibility of controlling page replacement as compared to algorithms A_ℓ^h that are designed only for programs occupying fixed memory sizes. The latter algorithms have the drawback that memory retains for an excessively long time those pages used relatively rarely, but whose use consists of several references in succession. For instance, let page x be referenced $h < m$ times in succession and from then onwards let it not appear in the reference string $\mathbf{x}$. After the h-th reference to page x algorithm A_ℓ^h assigns it the first position on list L_1, where it may remain for an indefinitely long time, although no longer needed. On the other hand, consider the self-organizing algorithm defined by the same parameter h and the parameters $T_1 = T_2 = ... = T_h = \infty$, $\mu_h = 2$ and $D_h = i_{h2}$. This algorithm replaces page x not more than two page faults after it has been introduced into memory.

3. Program Behavior Models and Performance Indexes

Deterministic and Stochastic Models. To compare various replacement algorithms it is first necessary to define the method of specifying reference strings. A simple such method assumes as given a fixed collection of reference strings obtained by recording the page references during execution of some set of programs. Thus, entire reference strings are assumed to be known in advance. The program behavior model constructed with this method of specifying reference strings is called deterministic. Within this deterministic framework replacement algorithms are compared over each individual reference string. Clearly, in order to obtain reliable comparisons it is necessary to use a sufficiently large number of reference strings. Moreover, there is the question of how representative the sample of reference strings is. The difficulty of evaluating representativeness and the excessive costs of generating reference strings are among the principal reasons for considering stochastic models of program behavior.

In a stochastic model, reference strings are considered to be realizations of a random process with a finite number of parameters not dependent on the lengths of reference strings. The simplest stochastic model of program behavior, namely the *independent reference model* (IRM), describes the string $x_1, x_2, ...,$ as a sequence of independent, identically distributed random variables with the probability distribution

(1) $Pr\{x_t = i\} = p_i$, $i = 1, 2, ..., n$; $t = 1, 2, ...$

It can be assumed that pages are numbered so that the probabilities satisfy $p_1 \geqslant p_2 \geqslant ... \geqslant p_n$.

The direct generalization of the IRM is the Markov model, which describes the reference string $x_1, x_2, ...$ by an ergodic, finite Markov chain (see [KemS]). For a set of pages $X = \{1, 2, ..., n\}$ the chain is defined by the transition probability matrix $\|p_{ij}\|_{i,j=1}^n$, where $p_{ij} = Pr\{x_t = j \mid x_{t-1} = i\}$. This model was first introduced in [MatGST].

As a special case we introduce a *partial* Markov model with a specific transition matrix $\|p_{ij}\|_{i,j=1}^n$ having elements

$$p_{ii} = \delta_i + (1-\delta_i)p_i, \quad i = 1, 2, ..., n,$$

(2)

$$p_{ij} = (1-\delta_i)p_j, \quad i, j = 1, 2, ..., n, \quad i \neq j,$$

where $0 \leqslant \delta_i \leqslant 1$, $0 < p_i < 1$, $\sum_{i=1}^n p_i = 1$. Such a model is interpreted as follows. At the initial time instant page i is chosen independently in accordance with the distribution $\{p_1, p_2, ..., p_n\}$ from the set X. If K_i denotes the total number of consecutive references to this page, then K_i is a random variable with the geometric distribution $Pr\{K_i = k\} = (1-\delta_i)\delta_i^{k-1}$, i.e. successive references to page i occur with probability δ_i. After the K_i references to page i the next page is again chosen independently from X in accordance with $\{p_1, p_2, ..., p_n\}$, and the process repeats. The distribution $\{p_1, p_2, ..., p_n\}$ and the probabilities δ_i, $i = 1, 2, ..., n$, are referred to as the parameters of the partial Markov model.

Generalization of the partial Markov model to an arbitrary distribution for K_i leads to the simplest semi-Markov model of program behavior. The only constraint is the requirement of a finite expectation $E(K_i) < \infty$, $i = 1, 2, ..., n$.

In the above models time is assumed to be discrete. However, for the analysis of certain replacement algorithms it is more convenient to use program behavior models in continuous time. The IRM in continuous time is given by the superposition of n independent Poisson processes with parameters $\beta_1, \beta_2, ..., \beta_n$. The Poisson process with parameter β_i determines the reference stream to page i, with $\sum_{i=1}^n \beta_i = 1$; therefore, the parameter β_i can be interpreted as the probability of referencing page i. In this model the intervals between two consecutive references to the same page have an exponential distribution. In the general case we replace the exponential distribution by an arbitrary one to obtain a renewal program-behavior model.

In particular, we have a renewal model [OpdC] if

(a) there are n distribution functions $G_i(x)$, $i = 1, 2, ..., n$ that describe n independent random variables, where the first moment $1/\lambda_i = \int_0^\infty [1-G_i(x)]dx$ exists for each i;

(b) the intervals between consecutive references to page i are independent and distributed in accordance with $G_i(x)$;

(c) the normalizing condition is satisfied; $\sum_{i=1}^n \lambda_i = 1$.

Since $1/\lambda_i$ is the expected time between consecutive references, condition (a) means that on the average there is one page reference per unit time.

Performance Indexes. In order to evaluate and compare the various replacement algorithms, it is necessary to introduce a performance index. Since the replacement algorithm influences primarily the frequency of page faults or, equivalently, the frequency of references to auxiliary memory, this measure should determine the performance index.

We define for any m-element subset $B \subset X$ an indicator function of its complement,

(3)
$$\phi(i,B) = \begin{cases} 1 & \text{if } i \notin B, \\ 0 & \text{if } i \in B. \end{cases}$$

We further assume that $|X_0| = m$, the memory size occupied by the program. For the deterministic model the performance index specifies the number of page faults induced by replacement algorithm A for some reference string of length N:

(4)
$$\Phi_m(A) = \sum_{t=1}^N \phi(x_t, X_{t-1}^A).$$

For all of the above stochastic models in discrete time the performance index is given in the form

(5)
$$F(A) \equiv F_{m,n}(A) = \lim_{N \to \infty} \sup \frac{1}{N} \sum_{t=1}^N \phi(x_t, X_{t-1}^A).$$

It should be noted that for each algorithm A considered later in this chapter, $\lim\limits_{N \to \infty} \sup (1/N) \sum\limits_{t=1}^{N} \phi(x_t, X_{t-1}^A)$ converges with probability one to $F(A) = \lim\limits_{t \to \infty} \sup E[\phi(x_t, X_{t-1}^A)]$ independently of the initial memory state X_0. In other words, averaging the number of faults per reference over the family of reference strings is asymptotically equivalent to finding the number of faults for one very long reference string.

The quantity $F(A)$ is referred to as the *page fault rate* or *miss ratio* of replacement algorithm A, and has the physical interpretation of being (with probability one) the long-term frequency of page faults for each reference string produced. The fact that the definition of criterion (5) contains a reference string of infinite length is not an excessive idealization; from operating system programs very long traces of page references can be recorded. Later, explicit expressions are obtained for criterion (5), and this facilitates considerably the comparison of replacement algorithms.

Model Adequacy. For any stochastic program-behavior model, the question arises as to how well it describes the traces of page references in real programs, i.e., the question of model adequacy or realism. Adequacy can be achieved by complicating the model, but this reduces the prospects of a successful analysis of replacement algorithms. Three notions of adequacy are introduced: strong, wide, and weak.

We say that there is adequacy in the *strong* sense if for the reference string of a real program the assumptions that define the model are satisfied with a given statistical confidence. Usually, in practical applications adequacy in the strong sense is not required. Normally, it is sufficient to achieve a specified proximity of values of a given functional defined over both the model and real reference strings. Under these circumstances we have adequacy in the *wide* sense.

In comparing replacement algorithms, it it is desirable to examine models representing not only individual properties of real sequences of page references, but also the properties of *locality* [DenK, MadB, SheT, SpiD] and of *rare references*. Consistent with our earlier notion of locality set, the locality property is exhibited when a program tends to use small subsets of its pages for relatively long periods of time. A program has the property of rare references if, in spite of locality sets, the reference strings contain rarely used pages. We say that there is adequacy in the *weak* sense if the model reflects the two properties of locality and rare references.

Attempts to construct a reasonable simple stochastic program behavior model adequate in the strong sense have so far not been successful. In [BogF] an example is given of adequacy in the wide sense of a Markov model. The difference between the LRU miss ratios computed for the model and for real page reference strings averaged 10%, and was computed for three different memory sizes assigned to the program.

The next section contains examples of adequacy in the wide sense of a partial Markov model and a renewal model. All the stochastic models examined here are adequate in the weak sense, because they reflect the above two properties of real reference strings (the partial Markov model and the simplest semi-Markov model reflect the simplest form of locality when a page is referenced several times in succession).

4. Examples of Adequate Models

The Case of Discrete Time [Eas]. Let $x_1, x_2, ..., x_t, ...$, be a page reference string of an n-page program whose pages comprise the set $X = \{1, 2, ..., n\}$. Formalizing our earlier discussion, we begin with

Definition 5. Let T be a positive integer. For each $t \geqslant T$ the working set $W(t, T)$ is the set of n distinct pages in the set $(x_{t-T+1}, ..., x_t)$. T is referred to as the *window size*, or parameter of the working set.[†] For $t > T$, let $S(t, T) = |W(t, T)|$ be the random variable denoting working set size.

We limit our consideration to stationary sequences $x_1, x_2, ..., x_t, ...$. This means that for any time instants $t_1, ..., t_k$, any elements $\{j_1, ..., j_k\} \subseteq X$, and any integer $h \geqslant 1$,

$$Pr\{x_{t_1} = j_1, ..., x_{t_k} = j_k\} = Pr\{x_{t_1+h} = j_1, ..., x_{t_k+h} = j_k\}.$$

---

† The continuous case is distinguished from the discrete one by our use of τ instead of T as the window size.

For each $t > T$ we define a random variable, $\Phi(t, T) \equiv \phi(x_{t+1}, W(t, T))$, denoting the page reference indicator of the working set (see (3)), and the expected working-set miss ratio

$$\zeta(t, T) = E[\Phi(t, T)] .$$

The stationarity of $\{x_t\}$ implies the stationarity of $\{\Phi(T, T), \Phi(T+1, T), ...\}$; therefore the expectation $\zeta(t, T)$ is not a function of t and is denoted $\zeta(T)$. The stationarity of $\{x_t\}$ also implies the stationarity of $\{S(T, T), S(T+1, T), ...\}$. Consequently the expected working-set size $\omega(T) = E[S(t, T)]$ is not a function of t. Since

$$S(t, T+1) = S(t, T) + \Phi(t, T),$$

then

(6)
$$\omega(T+1) = \omega(T) + \zeta(T) .$$

Since $S(1) = 1$, we have

$$\omega(T) = 1 + \sum_{k=j}^{T-1} \zeta(k) .$$

The values of $\zeta(T)$ and $\omega(T)$ may be estimated from a given reference string $\{x_t\}$. We define the random variables

(7)
$$\hat{\Phi}_j(T) = \frac{1}{j-T+1} \sum_{t=T}^{j} \Phi(t, T)$$

and

(8)
$$\hat{S}_j(T) = \frac{1}{j-T+1} \sum_{t=T}^{j} S(t, T) .$$

The values $\hat{\Phi}_j(T)$ and $\hat{S}_j(T)$ are unbiased estimates of $\zeta(T)$ and $\omega(T)$ respectively, i.e. $E[\hat{\Phi}_j(T)] = \zeta(T)$ and $E[\hat{S}_j(T)] = \omega(T)$. By the ergodic theorem [Yag] these estimates are consistent; i.e. they converge in probability as $j \to \infty$ to $\zeta(T)$ and $\omega(T)$ respectively, under the assumption that for any fixed T

$$\lim_{k \to \infty} E[\Phi(t, T)\Phi(t+k, T)] = [\zeta(T)]^2$$

and

$$\lim_{k \to \infty} E[S(t, T)S(t+k, T)] = [\omega(T)]^2 .$$

We now derive formulas for $\omega(T)$ and $\zeta(T)$ when the reference string $\{x_t\}$ is described by a partial Markov model with the parameters $\{p_1,..., p_n\}$ and $\delta_i \equiv \delta$. The transition probabilities of the Markov chain in this model are (see (2))

(9)
$$p_{ii} = \delta+(1-\delta)p_i ,$$
$$p_{ij} = (1-\delta)p_j, \quad i \neq j ,$$

where $0 \leqslant \delta \leqslant 1$, $\sum_{i=1}^{n} p_i = 1$ and $p_i > 0$ for $i = 1, 2,..., n$.

It is evident that $(p_1,..., p_n)$ is an invariant probability vector (stationary distribution) in such a Markov chain. Let x_1 be a random variable with the distribution $(p_1,..., p_n)$; the conditional distribution of x_{t+1} given the value of x_t is described by (9). Thus, $\{x_t\}$ is a stationary sequence.

We define the sequence of random variables

$$\chi_i(t, T) = \begin{cases} 1, & \text{if } x_j = i \text{ for some } j = t-T+1,...,t; \\ 0, & \text{otherwise .} \end{cases}$$

In these terms we obtain for the working-set size

$$S(t, T) = \sum_{i=1}^{n} \chi_i(t,T) .$$

By using the Markov property we have

$$E[\chi_i(t,T)] = 1-Pr\{x_j \neq i \text{ for } j = t-T+1,..., t\}$$
$$= 1-(1-p_i)[1-p_i(1-\delta)]^{T-1} .$$

Thus the following theorem is proved.

Theorem 1. For a partial Markov model described by a Markov chain with the transition probabilities (9), the expected working-set size is

(10)
$$\omega(T) = n - \sum_{i=1}^{n} (1-p_i)[1-p_i(1-\delta)]^{T-1} .$$

Corollary. By virtue of (6) the expected working-set miss ratio is

$$(11) \qquad \zeta(T) = (1-\delta) \sum_{i=1}^{n} p_i (1-p_i)[1-p_i(1-\delta)]^{T-1} .$$

Remark. The WS algorithm with parameter T retains in main memory at any time t only the pages of the working set $W(t, T)$. By the ergodic theorem it can be proved with the assumptions of Theorem 1 that

$$\lim_{T' \to \infty} \frac{1}{T'-T} \sum_{t=T}^{T'} \Phi(t, T)$$

converges with probability one to the quantity $F(\text{WS}) = F_T(\text{WS})$, the page fault rate of the WS algorithm, and that $F(\text{WS})$ equals the expected working-set miss ratio $\zeta(T)$. Thus, for the partial Markov program behavior model, $F(\text{WS})$ can be calculated from (11).

For the IRM (independent reference model) $\delta = 0$, so it follows from (10) and (11) that

$$(12) \qquad \omega(T) = n - \sum_{i=1}^{n} (1-p_i)^T ,$$

$$(13) \qquad \zeta(T) = \sum_{i=1}^{n} p_i (1-p_i)^T .$$

In approximating the working set characteristics, the partial Markov model examined here has been observed to describe the string of page references to a database used in a large-scale interactive administrative system of IBM (see [Wim]). We now consider the results of an analysis of such a sequence containing references to over 10^5 distinct pages. To test for adequacy, the theoretical and experimental values of the mean working-set size and miss ratio were calculated. Since in the present case adequacy in the strong sense does not hold, the parameter T was chosen relatively large $(T > 10^4)$ in order to accumulate a sufficient number of references to the individual program pages.

The probabilities $p_1 \geqslant p_2 \geqslant \ldots$ were estimated from observed relative frequencies. For smaller relative frequencies, viz. with 100 or fewer references to a page in a sample string, the power law extrapolation was used:

$$p_j = c/j^a ,$$

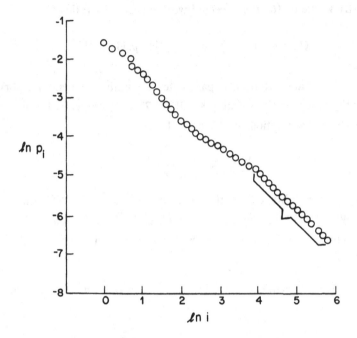

Figure 2 -   Estimates of p_i.

where $a = 0.96$ and $c = 0.08353$. Figure 2 shows estimates of the $\{p_i\}$ for the model. The value of n was determined from the normalization condition

$$\sum_{i=1}^{n} p_i = 1.$$

The expected working-set size $\omega(T)$ (Fig. 3) was computed using (10) for the $\{p_i\}$ values plotted in Fig. 2 and for $\delta = 0.79$. Figure 3 also shows the estimates $\hat{S}_j(T)$ computed from (8) for $j = 2 \times 10^6$. For comparison, the same figure shows the expected working-set sizes $\omega_{\text{IRM}}(T)$ for an IRM with the same probability distribution $\{p_i\}$. Figure 4 shows the expected working set miss ratios $\zeta(T)$ and their estimates $\hat{\Phi}_j(T)$ computed using (11) and (7), respectively, for the same parameters $\{p_i\}$, n, δ, and j. The graph of the expected working-set miss ratio $\zeta(T)$ for the corresponding IRM is not shown here, because most of its points differ considerably from the $\Phi(T)$ values.

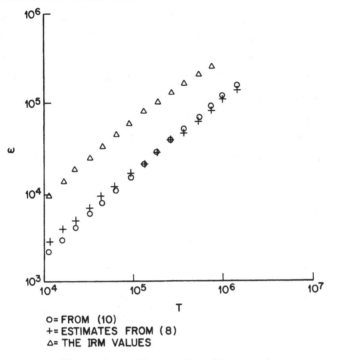

O= FROM (10)
+= ESTIMATES FROM (8)
△= THE IRM VALUES

Figure 3 -   Expected working set sizes.

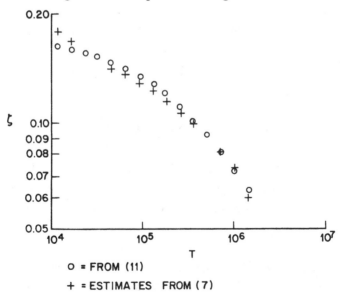

O = FROM (11)

+ = ESTIMATES FROM (7)

Figure 4 -   Expected working set miss ratios.

Figures 3 and 4 demonstrate the close agreement between the theoretical and experimental values for the mean working-set size and the mean working-set miss ratio. It can thus be concluded that in this example the partial Markov model is adequate in the wide sense. It by no means follows that the reference string considered has all the properties of a partial Markov model. For instance, the OPT algorithm, according to which the $m-1$ pages with the highest probabilities $p_1, p_2, ..., p_{m-1}$ are permanently retained in main memory, is optimal in this model in the sense of the performance index (5). However, it can be seen from Fig. 5

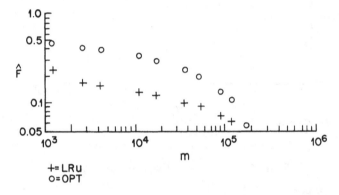

Figure 5 -   Estimates of the *LRU* and *OPT* miss ratios.

that the inequality $F_m(\text{LRU}) < F_m(\text{OPT})$ holds over a wide range of variation of the memory size used. Here

$$\hat{F}(A) - \hat{F}^{(N)}(A) - \frac{1}{N} \sum_{t-1}^{N} \phi(x_t, X_{t-1}^A),$$

where $\phi(.,.)$ is determined by (3) and N is fairly large, and where $\hat{F}(A)$ is an estimate of the miss ratio $F(A)$ (see (5)) for replacement algorithm A. In view of the strong law of large numbers for Markov chains $\hat{F}^{(N)}(A) \rightarrow F(A)$ with probability one as $N \rightarrow \infty$.

The Case of Continuous Time [OpdC]. We assume that time is measured only during program execution. This is referred to as virtual time. We assume that program behavior is described by the renewal model introduced in section 3.

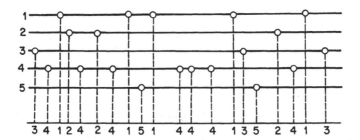

Figure 6 -   Example reference string generated by the renewal model.

Figure 6 shows an example of a reference string generated by this model. The horizontal axis represents virtual processing time. The reference sequence $\{x_t\}$ is formed by projecting the references of each page onto a common time axis. Extending Definition 5 to continuous time we have

Definition 6. Let $\tau > 0$. Then for each $t > \tau$ the working set $W(t, \tau)$ is the set of distinct pages in a segment of the reference string over the virtual time interval $(t-\tau, t]$, where τ is the parameter (window size) of the working set.

The page fault rate of the working set is

$$\zeta(\tau) = \lim_{t \to \infty} \frac{1}{t-\tau} \int_{\tau}^{t} \phi(x_t, W(s, \tau)) ds,$$

where $\phi(.,.)$ is determined by (3), and the equality holds with probability one.

Theorem 2. For the renewal model

(14)
$$\zeta(\tau) = \sum_{i=1}^{n} \lambda_i [1 - G_i(\tau)],$$

where $G_i(x)$ is the distribution function of the time intervals between two adjacent references to page i, and $1/\lambda_i$ is the expected length of these intervals, $i = 1, 2, ..., n$.

Proof. Applying the ergodic theorem and the law of total probability, we have

$$\zeta(\tau) = \sum_{i=1}^{n} P_i p_i$$

where

$$P_i = \lim_{t \to \infty} Pr\{i \notin W(t, \tau)\} \quad \text{and} \quad p_i = \lim_{t \to \infty} Pr\{x_t = i\} \ .$$

It is readily seen that $P_i = 1 - G_i(\tau)$, and $p_i = \lambda_i$, which yields (14). ∎

Remark. The WS algorithm with the continuous parameter τ retains in main memory at any time t only the pages of the working set $W(t, \tau)$. The definition of $\zeta(\tau)$ and Theorem 2 imply that the page fault rate of the WS algorithm can be calculated from (14).

Theorem 3. For the renewal model the expected working-set size $\omega(\tau) = E[S(t, \tau)]$ is not a function of t, and is given by

$$(15) \qquad\qquad \omega(\tau) = \sum_{i=1}^{n} \lambda_i \int_0^\tau [1 - G_i(x)] dx \ .$$

The proof of this theorem is based on the following lemma.

Lemma 1. The expected time spent by page i in the working set is

$$(16) \qquad\qquad \tau + \int_0^\tau x g_i(x) dx / (1 - G_i(\tau))$$

where $g_i(x) = dG_i(x)/dx$.

Proof. Page i is removed from the working set if it has not been referenced for the last τ time units. The probability that the first $k-1$ time intervals between successive references to page i do not exceed τ and that the k-th interval does exceed τ is $[G_i(\tau)]^{k-1}[1 - G_i(\tau)]$.

Let $E(Q_i|\tau)$ be the expected length of the inter-reference interval Q_i between two successive references to page i, given that it is shorter than τ. Then the expected time that page i will reside in main memory is

$$(17) \quad [1 - G_i(\tau)] \sum_{k=1}^{\infty} (k-1) E(Q_i|\tau) [G_i(\tau)]^{k-1} + \tau = \frac{E(Q_i|\tau) G_i(\tau)}{1 - G_i(\tau)} + \tau \ .$$

Since

$$Pr\{Q_i \leqslant x|\tau\} = \begin{cases} G_i(x)/G_i(\tau) & \text{if } x \leqslant \tau, \\ 0 & \text{if } x > \tau \end{cases}$$

then

(18)
$$E(Q_i|\tau) = \int_0^\tau xg_i(x)dx/G_i(\tau) .$$

Substituting (18) into (17) yields (16). ∎

Proof of Theorem 3. We say that there is a page fault caused by page i if $x_t = i$ and $i \notin W(t, \tau)$. The expected inter-page-fault interval for page faults caused by page i is $1/\lambda_i[1-G_i(\tau)]$ and is the reciprocal of the page fault rate created by page i. At the beginning of each such interval, page i belongs to the working set. The expected fraction of time page i resides in the working set is the ratio of the expected time page i resides in the working set to the expected inter-page-fault interval of page i, and is

$$\lambda_i \{\int_0^\tau xg_i(x)dx + \tau[1-G_i(\tau)]\} = \lambda_i \int_0^\tau [1-G_i(x)]dx .$$

Summing these time fractions over all pages of the program, we obtain (15). ∎

Differentiation of (15) yields

$$\zeta(\tau) = \frac{d}{d\tau} \omega(\tau)$$

which is a continuous analogue of (6). In other words, the page fault rate (the long-term frequency of references to pages not in the working set) equals the rate of variation of the expected size of the working set.

We now consider the results of testing the adequacy of the renewal model. In [OpdC] the sequences of references of two programs (programs 1 and 2) were analyzed. The distribution of the intervals between successive references to page i was approximated by the Weibull distribution

$$G_i(t) = 1-\exp\{-(\gamma_i t)^{\alpha_i}\}$$

Table 1

Page Number	$1/\lambda_i$	C_i	α_i	γ_i
1	7.2601	87.598	0.2077	12.085
2	11.601	60.762	0.2177	5.2498
3	68.537	43.704	0.2279	0.63832
4	158.78	31.026	0.2398	0.19592
5	46.714	20.631	0.2565	0.44153
6	148.94	28.481	0.2431	0.19146
7	322.69	21.901	0.2538	0.06799
8	1336.5	11.539	0.2861	0.0086463
9	1052.3	20.082	0.2576	0.019123
11	3.3813	239.34	0.1854	70.438
12	19.189	99.796	0.2045	5.1890
13	146.11	61.684	0.2173	0.42261
14	42.199	315.33	0.1802	7.4553
15	45.403	303.92	0.1809	6.6665
18	60.942	250.63	0.1844	4.1168
19	20.669	45.218	0.2268	2.1895
20	100.49	21.873	0.2540	0.21731
21	4.5785	45.907	0.2263	10.039
22	35.642	24.942	0.2483	0.70178
23	4544.0	28.834	0.2426	0.006358
25	26308.0	12.371	0.2822	0.0004698
26	41654.0	10.086	0.2943	0.0002426
32	2202.0	47.107	0.2254	0.02147

Table 2

Page Number	$1/\lambda_i$	C_i	α_i	γ_i
2	1741.8	12.878	0.2137	0.04025
3	166630.0	1.4573	0.7020	7.576E-6
4	31243.0	2.0555	0.5320	5.745E-5
5	14.113	124.34	0.1228	3887.45
6	74.6	53.605	0.1457	51.366
8	8.1907	12.625	0.2151	8.1422
9	8.8770	10.449	0.2296	4.6809
10	3.6379	41.857	0.1541	503.77
11	5.6674	34.610	0.1612	186.56
12	52.571	36.731	0.1590	23.694
13	90.315	21.083	0.1838	2.8669
14	298.27	15.949	0.1996	0.40926
15	54.591	36.331	0.1593	22.306
16	321.27	26.127	0.1734	1.4548
17	9.7798	43.766	0.1526	212.21
18	27.492	40.187	0.1555	59.515
20	278.49	16.401	0.1979	0.47184
21	31.394	6.9421	0.2687	0.50709

with the mean value

(19) $$1/\lambda_i = \Gamma(1+1/\alpha_i)/\gamma_i \ ,$$

and the coefficient of variation (ratio of the standard deviation to the mean)

(20) $$C_i = \left[\frac{\Gamma(1+2/\alpha_i)}{\{\Gamma(1+1/\alpha_i)\}^2} \right]^{1/2} .$$

An estimate of the parameter α_i was derived from the sample coefficient of variation (20). To find the estimate for the parameter γ_i, the estimates α_i and the sample mean were substituted into (19). The results are given in Tables 1 and 2, respectively, for programs 1 and 2. It should be noted that in the computations of estimates for the expectation $1/\lambda_i$ and the standard deviation, the virtual time was changed by the number of references; it was assumed that 1000 references to main memory occur during one millisecond of virtual time.

The results in these tables show great differences in the expected inter-reference intervals of different pages. For instance, for program 2 the smallest inter-reference interval is 3.6379 references and the largest inter-reference interval is 166,630 references. A similar statement is true for the coefficient of variation. This illustrates how different pages may exhibit radically different behavior.

Figures 7 and 8 show the results for the page fault rate per 1000 references for programs 1 and 2 (the FORTRAN and the FORTCOMP program, respectively) as a function of the working set parameter τ, measured in milliseconds. The dashed lines represent the observed results. The solid curves were computed using (14) for a renewal model and an IRM. The exponential distribution in the IRM can be viewed as the special case of the Weibull distribution for which $\alpha_i = 1$ and $\gamma_i = \lambda_i$. Figures 9 and 10 give the results of the mean working-set size as a function of τ for the same two sample programs. It can be seen from Figs. 7 - 10 that the renewal model describes the behavior of the programs better than the IRM.

5. Optimal Replacement Algorithms

By definition, optimal replacement algorithms minimize the mean frequency of references to auxiliary memory. The problem of finding optimal replacement algorithms in various program behavior models is of interest for two reasons. First, the structure of optimal algorithms suggests the structures to be included in the design of practical replacement algorithms. Secondly, knowing the optimal values of the performance index, it is possible to find (by simulation or analytically) to what extent a given practical algorithm differs from an optimal one, thus providing insight into the limits of applicability of the practical algorithm.

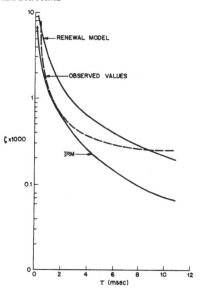

Figure 7 -   Theoretical and observed values of the working set miss ratio for program 1.

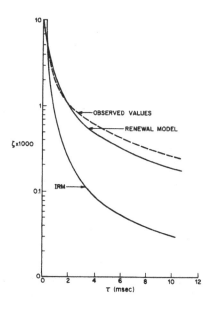

Figure 8 -   Theoretical and observed values of the working set miss ratio for program 2.

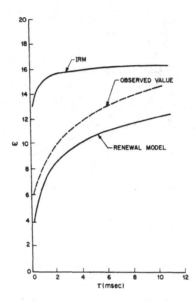

Figure 9 -   Theoretical and observed values of the working set size for
              program 1.

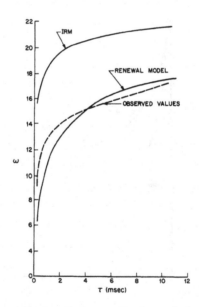

Figure 10 -   Theoretical and observed values of the working set size
              for program 2.

Later, we prove the optimality of the replacement algorithm A_0, which identifies the page y_t to be replaced at time t as the page having the largest expected time to the next reference among the pages in main memory. In particular, we show that A_0 is optimal in the class of demand replacement algorithms for the deterministic model with the performance index (4), and for the IRM, partial Markov and the simplest semi-Markov models with performance index (5).

In the general Markov model the optimal replacement algorithm A^* is obtained by solving a system of nonlinear algebraic equations and generally is not coincident with A_0. However, the results of a simulation show that for small programs algorithm A^* is very close to A_0 (in the sense of the performance index (5)).

In the deterministic model the entire reference string is known in advance. Therefore, the word "expected" can be dropped from the definition of algorithm A_0. The algorithm obtained is called the MIN algorithm [Bela] in order not to confuse the deterministic case with the stochastic one. A proof of the following result can be found in [CofD,MicS].

Theorem 4. For the deterministic program behavior model, the MIN algorithm minimizes the performance index (4) for any reference string x and fixed memory size m.

The proof of this result mirrors very closely the proof, to be given shortly, of the optimality of algorithm A_0 for the IRM of program behavior.

In the IRM the average time to the next reference to page i is $1/p_i$. According to algorithm A_0 the page $y = y_t^{A_0} \in X_{t-1}^{A_0}$ to be replaced is one for which the probability p_y is minimal among the pages in the set $X_{t-1}^{A_0}$ contained in main memory at time $t-1$. Since $p_1 \geqslant p_2 \geqslant ... \geqslant p_n$, then for any initial state X_0 of main memory the algorithm A_0 for any finite t leads to a main memory state $X_t \supset \{1, 2, ..., m-1\}$ with probability one. Thus, the algorithms A_0 and OPT, for which the page subset $\{1, 2, ..., m-1\}$ is permanently in main memory, give equal values of the performance index (5); i.e.

$$(21) \qquad\qquad F(A_0) = F(\text{OPT}) .$$

Now let

$$(22) \qquad f_k^A(B, t) = \sum_{r=1}^{k} E[\phi(x_{t+r}, X_{t+r-1}^A)] \,,$$

where $X_t^A = B$, and let

$$(23) \qquad f_k^*(B, t) = \min_A f_k^A(B, t) \,,$$

where A is a demand replacement algorithm.

Theorem 5. [AhoDU, Bro]. For the IRM defined by (1),

$$(24) \qquad f_k^{A_0}(B, t) = f_k^*(B, t)$$

for all $k > 1$, $B \subset X$ and $t \geqslant 1$. In other words, algorithm A_0 minimizes the expected number of page faults over any finite time interval.

This theorem has the following corollary.

Corollary. In an IRM algorithms A_0 and OPT are optimal in terms of the performance index (5).

Proof. The value of the index $f_k^*(B, t)$ satisfies the recurrence

$$(25) \quad f_k^*(B, t) = \sum_{i \in B} p_i f_{k-1}^*(B, t+1) + \sum_{i \notin B} p_i + \sum_{i \notin B} p_i \min_{z \in B} f_{k-1}(B+i\backslash z, t+1) \,,$$

which can be recognized as a dynamic programming formulation (the Bellman equation) of our page-fault minimization problem.

We introduce a linear order on the set X and say that a page $x \in X$ precedes page $y \in X$, written $x < y$, if $p_x < p_y$. If $p_x = p_y$, then $x < y$ if the number of page x is *larger* than the number of page y. To prove the theorem it is sufficient to show that for any $D \subset X$, $|D| = m-1$, and $x < y$, with $x, y \in X\backslash D$, we have the inequality

$$(26) \qquad f_k^*(D+x, t-1) - f_k^*(D+y, t-1) \geqslant 0 \,,$$

where $t \geqslant 1$.

The inequality (26) will be proved by induction on k. Applying (25) we can write

(27)
$$\Delta = f_k^*(D+x, t-1) - f_k^*(D+y, t-1) = \sum_{i \in D} p_i \Delta_1 +$$
$$+ \sum_{i \in D+x+y} p_i \Delta_2 + p_x(\Delta_3+\Delta_4) + (p_y-p_x)\Delta_4$$

where

$$\Delta_1 = f_{k-1}^*(D+x, t) - f_{k-1}^*(D+y, t),$$
$$\Delta_2 = \min_{z \in D+x} f_{k-1}^*(D+x \backslash z+i, t) - \min_{z \in D+y} f_{k-1}^*(D+y \backslash z+i, t),$$
$$\Delta_3 = f_{k-1}^*(D+x, t) - \min_{z \in D+y} f_{k-1}^*(D+y \backslash z+x, t) - 1,$$
and
$$\Delta_4 = \min_{z \in D+x} f_{k-1}^*(D+x \backslash z+y, t) - f_{k-1}^*(D+y, t) + 1.$$

Since $f_0^*(D+x, t) = f_0^*(D+y, t) = 0$, the inequality (26) holds for $k = 0$. Suppose it holds for all $0 \leqslant k < k'$; we show that it must then hold for $k = k'$.

Clearly, $0 \leqslant \Delta \leqslant 1$ by the inductive hypothesis. Let d be the least element in subset D. There are three cases to consider: $d < x < y$, $x < d < y$ and $x < y < d$. Accordingly, Δ_2 simplifies to

i) $\Delta_2 = f_{k-1}^*(D+x \backslash d+i, t) - f_{k-1}^*(D+y \backslash d+i, t), \quad d < x < y,$

ii) $\Delta_2 = f_{k-1}^*(D+i, t) - f_{k-1}^*(D+y \backslash d+i, t), \quad x < d < y,$

iii) $\Delta_2 = f_{k-1}^*(D+i, t) - f_{k-1}^*(D+i, t) = 0, \quad x < y < d.$

In the first two cases, on applying the inductive hypothesis we have $0 \leqslant \Delta_2 \leqslant 1$. The expression $(\Delta_3+\Delta_4)$ also simplifies to

$\Delta_3+\Delta_4 = f_{k-1}^*(D+x, t) - f_{k-1}^*(D+y \backslash d+x, t) + f_{k-1}^*(D+x \backslash d+y, t) - f_{k-1}^*(D+y, t)$

i) $= f_{k-1}^*(D+x, t) - f_{k-1}^*(D+y, t), \quad d < x < y,$

$\Delta_3+\Delta_4 = f_{k-1}^*(D+x, t) + f_{k-1}^*(D+y, t) - f_{k-1}^*(D+y \backslash d+x, t) - f_{k-1}^*(D+y, t)$

ii) $= f_{k-1}^*(D+x, t) - f_{k-1}^*(D+y \backslash d+x, t), \quad x < d < y,$

$\Delta_3+\Delta_4 = f_{k-1}^*(D+x, t) - f_{k-1}^*(D+x, t) + f_{k-1}^*(D+y, t) - f_{k-1}^*(D+y, t)$

iii) $= 0, \quad x < y < d.$

Applying the inductive hypothesis where necessary it can be shown that for each case $0 \leqslant \Delta_3+\Delta_4 \leqslant 1$.

It is convenient to rewrite Δ_4 in the form $\Delta_4 = 1 - \Delta_5$ where

$$\Delta_5 = f_{k-1}^*(D+y, t) - \min_{z \in D+x} f_{k-1}^*(D+y \backslash z + x, t) .$$

There are two cases to consider:

$$d < x => z = d => \Delta_5 = f_{k-1}^*(D+y, t) - f_{k-1}^*(D+x \backslash d+y, t) ;$$
$$x < d => z = x => \Delta_5 = f_{k-1}^*(D+y, t) - f_{k-1}^*(D+y, t) = 0 .$$

In both cases, $0 \leqslant \Delta_5 \leqslant 1$ and hence $0 \leqslant \Delta_4 \leqslant 1$. Using the upper and lower bounds on Δ_i, $i = 1, 2, 3, 4$, in (27) and noting that $p_y \geqslant p_x$, we find $0 \leqslant \Delta \leqslant 1$ as required. This proves inequality (26) and Theorem 5. ■

In the simplest semi-Markov and partial Markov models of program behavior, the expected inter-reference interval for page i, given that the page fault occurs on page j, equals $E(K_j) + 1/p_i$. Thus, according to algorithm A_0, page $y_t^{A_0} \in X_t^{A_0}$ is replaced with the minimum probability p_i. Using the correspondence between the semi-Markov and independent reference models of program behavior, we obtain the following theorem.

Theorem 6. In the simplest semi-Markov model of program behavior, which has the partial Markov model, (2), as a special case, algorithm A_0 is optimal for performance index (5).

Remark. As was the case in the IRM, in the simplest semi-Markov model, for any initial state X_0 of main memory, algorithm A_0 produces the memory state $X_t^{A_0} \supset \{1, 2, ..., m-1\}$ during a time interval that is finite with probability one, since $p_1 \geqslant p_2 \geqslant ... \geqslant p_n$. Consequently, algorithms A_0 and OPT, which retain the subset $\{1, 2, ..., m-1\}$ permanently in memory, have equal values of the performance index (5).

We conclude this section with a numerical procedure that leads to an optimal replacement algorithm A^* and an optimal value of the performance index

$$\xi = \inf_A F(A)$$

in a general ergodic Markov program behavior model [Kog1, BroK, AveK1, AveK2]. As previously, we consider only the class of demand replacement algorithms. A similar study of a program behavior model described by an absorbing Markov chain can be found in [IngK].

To find A^* and ξ we introduce an auxiliary, control Markov chain M^σ. The state space of M^σ is denoted by V and has elements (i, B) where i is any element of B and B is any subset of X with $|B| = m$. Let $\tau(B) = \inf \{t: x_t \in B\}$ and

$$\mu(i, B) = E[\tau(B)|x_0 = i]$$
$$q(i, B, \ell) = Pr\{x_{\tau(B)} = \ell | x_0 = i\} .$$

The values of $\mu(i, B)$ and $q(i, B, \ell)$ can be computed from the formulas in [KemS], in terms of the transition probability matrix $\|p_{ij}\|_{i,j=1}^n$ of the initial Markov chain M describing program behavior. Thus, the transition probabilities of the chain M^σ with $\sigma = \sigma(A)$ corresponding to replacement algorithm A are given by

$$
\begin{aligned}
Pr\{(x_{s+1}, B_{s+1}) &= (\ell, D)|(x_s, B_s) = (i, B)\} \\
&= 0 , \quad \text{if } \ell \in B , \\
&= q(i, B, \ell)\delta(D, B\backslash y^A + \ell), \quad \text{if } \ell \notin B ,
\end{aligned}
$$

where (ℓ, D) and (i, B) are in V, algorithm A replaces page $y^A = y^A(i, B)$ and

$$\delta(D, B\backslash y + \ell) = \begin{cases} 1 & \text{if } D = B\backslash y + \ell \\ 0 & \text{otherwise .} \end{cases}$$

Suppose a control σ is given. Then the expected return from one step starting in state (i, B) is equal to $\mu(i, B)$, and the average return per unit time is equal to

$$f^\sigma(i, B) = \lim_{T \to \infty} \inf \frac{1}{T} \sum_{s=1}^T E[\mu(x_s, B_s)|(x_0, B_0) = (i, B)] .$$

The optimal control σ^* should maximize $f^\sigma(i, B)$. We set

$$\theta = \sup_\sigma f^\sigma(i, B) .$$

Lemma 2. Let the control chain M^σ with performance index f^σ be given. Then from any stationary Markovian control σ' whose corresponding Markov chain $M^{\sigma'}$ has more than one ergodic class, we can construct a stationary Markovian control β such that the Markov chain M^β has one ergodic class, and $f^\beta(i, B) \geqslant f^{\sigma'}(i, B)$ for all $(i, B) \in V$, where $f^\beta(i, B) \equiv c$ is not a function of (i, B).

Proof. Using the algorithm proposed in [FoxL] we can isolate the ergodic class U with maximal value $f^{\sigma'}(i, B)$. We have $f^{\sigma'}(i, B) = c$ for $(i, B) \in U$, and because of our assumption, there exists $(i, B) \in V \setminus U$ for which $f^{\sigma'}(i, B) \leqslant c$. Let $(j, D) \in U$ and let β be a stationary Markovian control which coincides with σ' for $(i, B) \in U$, while for $(i, B) \in V \setminus U$ it corresponds to a replacement algorithm A_β for which $y_t^{A_\beta} \in X_{t-1} \setminus (D \setminus j)$, where $X_{t-1} = B$. Since the Markov chain M is ergodic, we have $f^\beta(i, B) = c$ for any $(i, B) \in V$. ∎

Corollary. Assume that U^* is an arbitrary ergodic class of the chain M^{σ^*}. Then

$$(28) \qquad \theta = \sum_{(i, B) \in U^*} \pi^*(i, B) \mu(i, B)$$

where $\{\pi^*(i, B)\}$ is a fixed probability vector of the reduced Markov chain M^{σ^*} with state space U^*.[†]

Theorem 7. For a Markov model of program behavior given by the ergodic transition probability matrix $\|p_{ij}\|_{i, j=1}^n$, the relation

$$(29) \qquad \xi = 1/\theta$$

holds. The value of θ is uniquely determined by the system of equations

$$(30) \qquad \theta + v(i,B) = \mu(i, B) + \sum_{\ell \notin B} q(i, B, \ell) \max_{j \in B} v(\ell, B \setminus j + \ell)$$

with the additional unknowns $v(i, B)$, where (i, B) runs through the entire set V. If $x_t \notin X_{t-1}$ and $x_t = i$, then the optimal algorithm A^* replaces that page y_t^* for which

$$v(i, X_{t-1} \setminus y_t^* + i) = \max_{j \in X_{t-1}} v(i, X_{t-1} \setminus j + i) .$$

The proof of the theorem follows from the definition of the chain M^σ, Lemma 2 and the corresponding results from the theory of controlled Markov chains [Der].

Several remarks can be made about the numerical procedure for determining the optimal replacement algorithm A^* and the optimal performance criterion ξ.

---

† The elements of the transition matrix of this chain are those elements of the transition matrix of M^{σ^*} which correspond to the set U^*.

1. By Lemma 2 a modification of Howard's method of policy iteration [How] can be used to find A^* and ξ. If an iteration leads to a control σ' for which $M^{\sigma'}$ has more than one ergodic class, then by virtue of Lemma 2 the control σ can be replaced by a control σ_1 for which M^{σ_1} has one ergodic class.

2. Equations (30) are especially simple when $n - m = 1$. In this case $q(i,B,\ell) \equiv 1$, and from (28) and (29) it follows that

$$\xi = \left[\sum_{(i,B) \in U} \mu(i, B)/L \right]^{-1},$$

where L is the number of elements in ergodic class U, and $\mu(i, B)$ is equal to $m_{i\ell}$, the expected first passage time of state $\ell \notin B$ starting in state i in chain M.

3. In each equation of the system (30) only $n-m+3$ coefficients are nonzero. The number N of equations in (30) decreases sharply with increases in the number of pages stored permanently in memory. Indeed, if there are $k \leqslant m - 1$ such pages, then

$$N = (m - k) \binom{n - k}{m - k} .$$

A condition sufficient for the pages of a set $Y \subset X$ to be kept permanently in memory by an optimal algorithm is given in the following result.

Theorem 8. Let $Y = \{1, 2, ..., k\}$ and

$$r = \min_{\substack{1 \leqslant i \leqslant n \\ k+1 \leqslant j_1, j_2 \leqslant n}} (p_{ij_1}/p_{ij_2}) ,$$

where in calculating the minimum it is assumed that $0/0 = 1$. Assume that $r > 0$ and for all $i, j \neq i$ and $1 \leqslant j \leqslant k$, the inequalities

(31) $$rp_{ij} \geqslant p_{i\ell}, \quad k + 1 \leqslant \ell \leqslant n$$

hold. Then there exists an optimal replacement algorithm A^* by which the pages of Y are never replaced.

Remark. Condition (31) is fairly limiting. It may, for instance, not be satisfied when M is a process of independent trials. However, an example is easily found which shows that, generally speaking, (31) can not be relaxed to the condition $p_{ij} \geqslant p_{i\ell}$.

Proof. It can be proved that ξ is uniquely determined by the system of equations

(32)
$$\xi + u(i, B) = \sum_{j \notin B} p_{ij} + \sum_{j \in B} p_{ij} u(j, B) + \sum_{j \notin B} p_{ij} \min_{\ell \in B} u(j, B \setminus \ell + j),$$

with the additional unknowns $u(i,B)$, where (i,B) runs through the entire set V.

Let $D \subset X$ be such that $|D| = m - 1$ and $|Y \setminus D| = 1$. To prove the theorem it is sufficient to show that

(33)
$$\min_{i, j, \ell, D} [u(i, D+\ell) - u(i, D+j)] \geqslant 0,$$

where the minimum is taken over all admissible sets D and where $i \in D$, $j \in Y$, $\ell \in X \setminus Y$, and $\ell, j \in D$. Inequality (33) is proved by means of relations for the difference $[u(i, D + \ell) - u(i, D + j)]$ derived from (32). The details are routine and therefore omitted. ∎

We should note that in contrast to (30) the number of equations in (32) does not decrease with increases in the number, k, of pages permanently resident in main memory. Each equation in (32) has $n+3$ non-zero coefficients; the number of equations is $n \binom{n}{m}$, independent of k.

A question of some interest is the extent to which algorithm A_0 is worse than an optimal algorithm in the sense of performance index (5). To get an idea of this discrepancy we computed the ratio $F(A_0)/F(A^*)$ for 3000 4×4 matrices, $\|p_{ij}\|_{i,j=1}^4$, generated as follows. The three elements p_{i1}, p_{i2} and p_{i3} were sampled independently from the uniform distribution on the tetrahedron $\{(z_1, z_2, z_3) | 0 \leqslant z_1+z_2+z_3 \leqslant 1\}$, and the fourth element was taken as $p_{i4} = 1 - (p_{i1}+p_{i2}+p_{i3})$, $1 \leqslant i \leqslant 4$.

Recall that on a fault to page $x_t = i$, A_0 replaces a page j such that $m_{ij} = \max_{k \in X_{t-1}^{A_0}} m_{ik}$, where m_{ij} is the expected first passage time from state i to state j. Thus, for each $\|p_{ij}\|_{i,j=1}^4$ we calculated the matrix $\|m_{ij}\|_{i,j=1}^4$ from which

$F(A_0)$ was computed for $m = 2$ and 3. Results were obtained for an optimal algorithm A^* derived from Howard's method of policy iteration. The initial control σ_0 was selected as the control corresponding to algorithm A_0, and $F(A^*)$ was computed for $m = 2$ and 3. The results can be summarized as follows.

1. $A^* = A_0$ in 47% of the events for $m = 2$ and in 72% of the events for $m = 3$.

2. The iterative procedure converged to the solution in one to three iterations.

3. The maximum value of $F(A_0)/F(A^*)$ was 1.48 for $m = 3$ and the matrix

$$\|p_{ij}\| = \begin{vmatrix} 0.231 & 0.068 & 0.688 & 0.013 \\ 0.223 & 0.048 & 0.039 & 0.690 \\ 0.392 & 0.076 & 0.501 & 0.031 \\ 0.351 & 0.479 & 0.120 & 0.050 \end{vmatrix}$$

For this matrix A_0 does not replace pages 1 and 3 and A^* does not replace 1 and 2. It is interesting to note that in this case A_0 is close to LRU ($F(\text{LRU}) = 0.160$ and $F(A_0) = 0.166$), while A^* is close to FIFO ($F(\text{FIFO}) = 0.119$ and $F(A^*) = 0.113$).

6. Expressions for Miss Ratios

General Method. We consider the class of demand replacement algorithms with finite memory that determine replaced pages y_t^A by functions only of the state of main memory X_{t-1}^A, i.e. $y_t^A = f(X_{t-1}^A)$, where the rule f may be deterministic or randomized. This class contains the subclass of replacement algorithms that apply when fixed memory sizes are assigned to programs. RR and OPT are among the algorithms in this class.

For an independent reference model of program behavior we have the following general method of evaluating the miss ratio $F(A)$ defined by (5). When the references $x_1, x_2,...$ are independent, the evolution of the memory states $\{X_t^A\}$ is described by a finite Markov chain M^A on the set of states $\{s\}^A$, whose elements are the ordered sets of m distinct pages $(i_1, i_2,..., i_m) \in X$ [Kin]. Let $c^A = \|c(s, s')\|$ be the transition probability matrix for this chain. It can be shown that such chains are ergodic for the replacement algorithms A considered below. Thus a unique invariant probability vector $\mathbf{q}^A$ exists such that $\mathbf{q}^A c^A = \mathbf{q}^A$ [KemS]. We denote the elements of this vector by $q(s)$, where s is a state of M^A. The ergodic theorem implies that the miss ratio $F(A)$ is equal to the stationary probability of page faults, which is obtained by the formula of total probability as

$$(34) \qquad F(A) = \sum_s q(s) \sum_{s'}{}' c(s, s') ,$$

where summation in the outer sum is with respect to all states of M^A and $\sum'$ denotes a summation only over those elements s' satisfying the condition $|s'\backslash s| = 1$, i.e. differing from s in precisely one page. Therefore, evaluation of $F(A)$ amounts to calculating $\mathbf{q}^A$. In general, this vector can only be found numerically. However, in the IRM explicit expressions for $\mathbf{q}^A$, and therefore $F(A)$, are derived for various replacement algorithms. One of these results can then be generalized to the partial Markov program behavior model. We now consider the IRM in more detail.

The IRM. We apply the general method to the analysis of a number of the important replacement rules. The basic steps of the method are illustrated, but the details are left to the interested reader.

Theorem 9. [Kin]. Under the IRM we have

$$(35) \quad F_m(\text{LRU}) = \sum_{(i_1,\dots,\, i_m)\, \in\, \Lambda_{m,\, n}} \frac{p_{i_1}\cdots p_{i_m} \sum_{j=m+1}^{n} p_{i_j}}{(1-p_{i_1})(1-p_{i_1}-p_{i_2})\dots(1-p_{i_1}-p_{i_2}-\dots-p_{i_{m-1}})} ,$$

where the outer sum is over the set $\Lambda_{m,\, n}$ of all sequences (permutations) of distinct elements taken m at a time from $\{1, 2,\dots, n\}$.

Proof. Following the general method we construct an auxiliary Markov chain M^{LRU} on the set $\Lambda_{m,\, n}$. From state $s = (i_1,\dots, i_m)$ it is possible to reach in one step only states of the form $(i_\ell, i_1,\dots, i_{\ell-1}, i_{\ell+1},\dots, i_m)$ and $(i_r, i_1,\dots, i_{m-1})$, $i_r \notin s$. The respective probabilities are p_{i_ℓ}, $\ell = 2, 3,\dots, m$, and p_{i_r}, $r = m+1,\dots, n$. The probability of remaining in state s is p_{i_1}. We perform the necessary calculations and find that in this case the solution to the system of linear algebraic equations for the stationary probabilities is

$$
(36) \qquad q^{\text{LRU}}(s) = \frac{\prod_{k=1}^{m} p_{i_k}}{(1-p_{i_1})(1-p_{i_1}-p_{i_2})\dots(1-p_{i_1}-\dots-p_{i_{m-1}})}
$$

where $s = (i_1, i_2,\dots, i_m)$. Now (35) is implied by (34) and (36) and, hence the theorem is proved.

Theorem 10. [AveS, Kin]. Under the assumptions of Theorem 9

$$
(37) \qquad F_m(\text{FIFO}) = \frac{\sum\limits_{(i_1,\dots,\, i_m)\in\Lambda_{m,\,n}} p_{i_1}p_{i_2}\dots p_{i_m} \sum\limits_{j=m+1}^{n} p_{i_j}}{\sum\limits_{(i_1,\dots,\, i_m)\in\Lambda_{m,\,n}} p_{i_1}p_{i_2}\dots p_{i_m}} .
$$

Proof. The auxiliary chain M^{FIFO} is defined on the same set of states as M^{LRU}. Let $s = (i_1,\dots, i_m)$. The elements $c(s, s')$ of matrix c^{FIFO} differ from zero only where $s' = s$ and $s' = (i_\ell, i_1,\dots, i_{m-1})$ for $\ell = m+1, m+2,\dots, n$, and are equal to $\sum\limits_{r=1}^{m} p_{i_r}$ and p_{i_ℓ}, respectively. Consequently, for $n > m+1$ we have that

$$
(38) \qquad q^{\text{FIFO}}(s) = \frac{p_{i_1}p_{i_2}\dots p_{i_m}}{\sum\limits_{(i_1,\dots,\, i_m)\in\Lambda_{m,\,n}} p_{i_1}p_{i_2}\dots p_{i_m}} .
$$

If $n = m+1$, the ergodic set for M^{FIFO} consists of $m+1$ elements and is dependent on the initial state $s_0 = X_0$. It can be shown that for any ergodic set

$$
(39) \qquad q^{\text{FIFO}}(s) = \frac{p_{i_1}p_{i_2}\dots p_{i_m}}{\sum\limits_{\{i_1,\dots,\, i_m\}\in\Lambda'_{m,\,n}} p_{i_1}p_{i_2}\dots p_{i_m}} , \quad n = m+1 ,
$$

where $\Lambda'_{m,\,n}$ denotes the set of all combinations of the elements of X taken m at a time. Note that the elements of $\Lambda'_{m,\,n}$ are subsets of X while the elements of $\Lambda_{m,\,n}$ are *ordered* subsets of X. It follows that

$$
(40) \qquad \sum\limits_{(i_1,\dots,\, i_m)\in\Lambda_{m,\,n}} p_{i_1}p_{i_2}\dots p_{i_m} = m! \sum\limits_{\{i_1,\dots,\, i_m\}\in\Lambda'_{m,\,n}} p_{i_1}p_{i_2}\dots p_{i_m} .
$$

Thus, the statement of the theorem is implied by (34), (38) and (39).

Theorem 11. Under the assumptions of Theorem 9

$$
F_m(\text{RR}) = \frac{\displaystyle\sum_{\{i_1,\dots,\,i_m\}\in\Lambda'_{m,n}} p_{i_1}p_{i_2}\cdots p_{i_m}\sum_{j=m+1}^{n} p_{i_j}}{\displaystyle\sum_{\{i_1,\dots,\,i_m\}\in\Lambda'_{m,n}} p_{i_1}p_{i_2}\cdots p_{i_m}}.
$$

Proof. The auxiliary chain M^{RR} is defined on the set of states, $\Lambda'_{m,n}$. From state $\{i_1,\dots,i_m\}$ it is possible to reach in one step only states $\{i_\ell, i_2,\dots, i_m\}$, $\{i_1, i_\ell, i_2,\dots, i_m\},\dots, \{i_1,\dots, i_{m-1}, i_\ell\}$ with probabilities p_{i_ℓ}/m, for $\ell = m+1,\dots, n$. The probability of remaining in state s is $\sum_{r=1}^{m} p_{i_r}$. It is readily verified that

$$
q^{\text{RR}}(s) = \frac{p_{i_1}p_{i_2}\cdots p_{i_m}}{\displaystyle\sum_{\{i_1,\dots,\,i_m\}\in\Lambda'_{m,n}} p_{i_1}p_{i_2}\cdots p_{i_m}},
$$

and that the result of the theorem is implied by (34).

Remark. It follows from (40) that

$$
(41)\qquad F_m(\text{FIFO}) = F_m(\text{RR}) = (m+1)\frac{\displaystyle\sum_{\{i_1,\dots,\,i_{m+1}\}\in\Lambda'_{m+1,n}} p_{i_1}p_{i_2}\cdots p_{i_{m+1}}}{\displaystyle\sum_{\{i_1,\dots,\,i_m\}\in\Lambda'_{m,n}} p_{i_1}p_{i_2}\cdots p_{i_m}}.
$$

The following formulas can also be proved by the general method [AveBK1, AveBK2].

$$
F_m(A_\ell^1) = \sum_{\{i_1,\dots,\,i_m\}\in\Lambda_{m,n}} p_{i_1}p_{i_2}\cdots p_{i_m}\sum_{j=m+1}^{n} p_{i_j}/[\ell!(1-p_{i_1}-\dots-p_{i_\ell})\dots
$$
$$
(1-p_{i_1}-\dots-p_{i_{m-1}})\sum_{\{i_1,\dots,\,i_\ell\}\in\Lambda'_{\ell,n}} p_{i_1}p_{i_2}\cdots p_{i_\ell}],
$$

and

$$
(42)\ F_m(A_{m_1}^{m-m_1+1}) = \frac{\displaystyle\sum_{\{i_1,\dots,\,i_m\}\in\Lambda_{m,n}} p_{i_1}^{m-m_1+1}\cdots p_{i_{m_1}}^{m-m_1+1}p_{i_{m_1+1}}^{m-m_1}\cdots p_{i_m}\sum_{j=m+1}^{n} p_{i_j}}{\displaystyle\sum_{\{i_1,\dots,\,i_m\}\in\Lambda_{m,n}} p_{i_1}^{m-m_1+1}\cdots p_{i_{m_1}}^{m-m_1+1}p_{i_{m_1+1}}^{m-m_1}\cdots p_{i_m}}.
$$

The algorithms $A_{m_1}^{m-m_1+1}$ are intermediate between FIFO and CLIMB, which are obtained with the respective extreme values $m_1 = m$ and $m_1 = 1$. In particular, for $m_1 = m-1$ we get only two priority subsets: L_1, which contains $m-1$ elements, and L_2 which consists of one element. The set L_1 is updated in accordance with the FIFO algorithm only on a reference to a page whose number is contained in L_2. The set L_2 is updated as in the last priority subset L_m of CLIMB.

The following are the stationary probabilities of the auxiliary Markov chains for $A' = A_\ell^1$ and $A'' = A_{m_1}^{m-m_1+1}$, with $s = (i_1,\ldots, i_m)$:

$$q^{A'}(s) = \frac{p_{i_1} p_{i_2} \cdots p_{i_m}}{\ell!(1-p_{i_1}-\ldots-p_{i_\ell})\ldots(1-p_{i_1}-\ldots-p_{i_{m-1}}) \displaystyle\sum_{(i_1,\ldots,\, i_\ell) \in \Delta'_{\ell,\, n}} p_{i_1} p_{i_2} \cdots p_{i_\ell}},$$

and

$$(43) \qquad q^{A''}(s) = \frac{p_{i_1}^{m-m_1+1} \cdots p_{i_{m_1}}^{m-m_1+1} p_{i_{m_1+1}}^{m-m_1} \cdots p_{i_m}}{\displaystyle\sum_{(i_1,\ldots,\, i_m) \in \Delta_{m,\, n}} p_{i_1}^{m-m_1+1} \cdots p_{i_{m_1}}^{m-m_1+1} p_{i_{m_1+1}}^{m-m_1} \cdots p_{i_m}}.$$

Explicit expressions for $F_m(A)$ can be derived in the IRM for numerous other fixed-memory algorithms. For example, for the algorithm A_c defined by the parameters $h = 2$, $T_1 = T_2 = \infty$, $\mu_1 = \ell_1 = c$, $\mu_2 = \ell_2 = c$ and $D_2 = i_{21}$ the miss ratio is

$$F_m(A_c) = \sum_{(i_1,\ldots,\, i_m) \in \Delta_{m,\, n}} p_{i_1}^2 \cdots p_{i_c}^2 p_{i_{c+1}} \cdots p_{i_m} \sum_{j=m+1} p_{i_j} \Big/ \sum_{(i_1,\ldots,\, i_m) \in \Delta_{m,\, n}} p_{i_1}^2 \cdots p_{i_c}^2 p_{i_{c+1}} \cdots p_{i_m}.$$

Each replacement algorithm A may be associated with a so-called *partially preloaded* replacement algorithm RA that permanently maintains a certain subset of the pages in the memory.

Definition 7. Let R_k be any k-element subset of X ($0 \leqslant k \leqslant m-1$), with R_0 the empty set and $n > m > 0$. Let a replacement algorithm A be defined for all $0 < m < n$. The partially preloaded replacement algorithm RA corresponding to A and R_k is defined so that for any $t \geqslant 0$ we have $R_k \subset X_t^{RA}$ and RA acts as A with program pages $X \backslash R_k$, exactly $m-k$ of which can be kept in memory.

The following theorem establishes a relation between the miss ratios for algorithms A and RA.

Theorem 12. Let $\tilde{X} = X \backslash R_k = (i_{k+1}, i_{k+2},..., i_n)$ be the set of pages in a program assigned a memory size of $m-k$ pages. Let the references $x_1, x_2,...$ be a sequence of independent identically distributed random variables such that for all t

$$Pr\{x_t = i_r\} = \tilde{p}_{i_r} = p_{i_r}/(1-\beta_k) , \quad k+1 \leqslant r \leqslant n ,$$

where $\beta_k = \sum_{r=1}^{k} p_{i_r}$, $k \geqslant 1$, and $\beta_0 = 0$. For algorithm A let $\tilde{q}^A = (\tilde{q}^A(\tilde{s}))$ be an invariant probability vector for the auxiliary Markov chain $\tilde{M}^A$ that describes the evolution of the memory states $\tilde{X}_t^A$. Then for $R_k = (i_1, i_2,..., i_k)$ and $\tilde{s}$ a state of $\tilde{M}^A$,

$$q^{RA}(s) = q^{RA}(i_1, i_2,..., i_k, \tilde{s}) = \tilde{q}^A(\tilde{s}) ,$$

and

$$F_{m-k, n-k}(A) = F_{m, n}(RA)/(1-\beta_k) .$$

The proof follows directly from the conditions of the theorem, the definitions of q^A and RA, and from (34).

By Theorem 12 we obtain in particular that [BogK, AveK1]

(44)

$$F(RLRU) = \sum_{(i_{k+1},..., i_m) \in \Lambda_{m, n}(k)} \frac{p_{i_{k+1}} p_{i_{k+2}} \cdots p_{i_m} \sum_{j=m+1}^{n} p_{i_j}}{(1-\beta_k)(1-\beta_k-p_{i_{k+1}}) \cdots (1-\beta_k-p_{i_{k+1}}-\cdots-p_{i_{m-1}})} ,$$

where $\Lambda_{m, n}(k)$ is the set of all permutations of elements taken $m-k$ at a time from $\{i_{k+1},...,i_n\}$. Also, we have that [Gel1]

$$F_m(RFIFO) = F_m(RRR) = (m-k+1) \frac{\sum_{(i_{k+1},..., i_{m+1}) \in \Lambda'_{m+1, n}(k)} p_{i_{k+1}} p_{i_{k+2}} \cdots p_{i_{m+1}}}{\sum_{(i_{k+1},..., i_m) \in \Lambda'_{m, n}(k)} p_{i_{k+1}} p_{i_{k+2}} \cdots p_{i_m}} ,$$

where $\Lambda'_{m, n}(k)$ is the set of all subsets of $m-k$ elements taken from $\{i_{k+1},...,i_n\}$.

The following theorem gives a useful representation of F_m (LRU).

Theorem 13. Let $R(i)A$ be a partially preloaded replacement algorithm with $R_1 = \{i\}$. Then

(45) $$F_m(\text{LRU}) = \sum_{i=1}^{n} p_i F_m(R(i)\text{LRU}) .$$

The proof is derived from (35) and (44). To calculate (5) with the OPT algorithm, which is identical to the RLRU algorithm when the resident set is $R_{m-1} = \{1, 2,..., m-1\}$, we use (44) and get [AhoDU]

(46) $$F_m(\text{OPT}) = 2 \sum_{m \leqslant i_1 < i_2 \leqslant n} p_{i_1} p_{i_2} / \sum_{i=m}^{n} p_i - \sum_{i=m}^{n} p_i - \sum_{i=m}^{n} p_i^2 / \sum_{i=m}^{n} p_i$$

Comparison of Algorithms. We have a natural conjecture that the algorithms A_ℓ^h are partially ordered under the independent reference model as follows. First,

$$F(A_1^h(m_1,..., m_h)) \geqslant F(A_1^{h+1}(m_1',..., m_{h+1}')) ,$$

where for some $1 \leqslant r \leqslant h$ we have $m_i' = m_i$ for $1 \leqslant i \leqslant r-1$, $m_i' = m_{i-1}$ for $r+2 \leqslant i \leqslant h+1$, and $m_r' + m_{r+1}' = m_r$. Also, if $\ell_1 > \ell_2$ then

$$F(A_{\ell_1}^h(m_1,..., m_h)) \geqslant F(A_{\ell_2}^h(m_1,..., m_h)) ,$$

and finally,

$$\min_{\{m_i\}, \ell} F(A_\ell^h(m_1,..., m_h)) = F(A_1^h(m-h+1, 1,..., 1)) .$$

This means that

$$F(\text{FIFO}) \geqslant F(\text{LRU}) \geqslant F(\text{CLIMB})$$

and that

$$F(\text{FIFO}) = \max_A F(A) , \quad F(\text{CLIMB}) = \min_A F(A) ,$$

where the minimum and maximum are taken over the set of finite-memory demand replacement algorithms. Thus, CLIMB is conjectured to be optimal within the

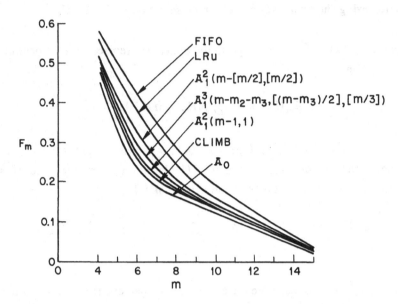

Figure 11 -   Miss ratio vs. memory size for A_0 and several A_1^h.

IRM. A similar assertion is actually proved in [KanR] for a restricted case when $p_1 = p, p_i = (1-p)/(n-1), 1 < i \leqslant n$.

Figure 11 shows the miss ratio as a function of the memory size m assigned to the program for several algorithms A_ℓ^h and A_0. Here, $n = 16$ and $p_1 = p_2 = \ldots = p_5 = 0.15$; $p_6 = 0.05$; $p_7 = p_8 = \ldots = p_{16} = 0.02$. Figure 12 compares simulation data for three algorithms A_ℓ^h with data for the MIN algorithm. The reference string used was a record of 12,000 references to elements (pages) in a partitioned data-set directory of operating system J [Wor]. The measurements were performed over 2.5 hours of normal operation of an ICL System 4-70 [AveK3].

The broken line $F_m(A_1^3)$ in Fig. 12 gives the best results among all possible decompositions $m = m_1+m_2+m_3$ for each value of memory size m. Elementary statistical analysis of the reference string shows that it cannot be described by an IRM. However, the FIFO, LRU and A_3^1 algorithms are ordered as in the IRM. The figure shows that

$$0.25 = F_{24}(A_3^1) = F_{32}(\text{LRU}) < F_{40}(\text{FIFO})$$

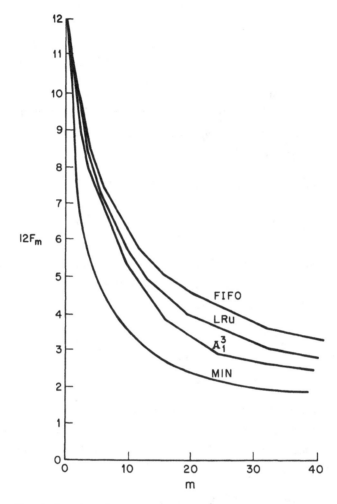

Figure 12 -   Simulation data for *FIFO, LRU, A_1^3* and *MIN*.

which shows that considerable economy in memory size can be gained by properly choosing the replacement algorithm.

Partial Markov and semi-Markov Models [Kog2, Kog4].   It can be shown for the semi-Markov model that

$$(47) \qquad F_m(\text{OPT}) = F_m^0(\text{OPT}) / \sum_{i=1}^{n} p_i \nu_i$$

and

$$(48) \qquad F_m(\text{LRU}) = F_m^0(\text{LRU})/\sum_{i=1}^{n} p_i \nu_i$$

where $\nu_i = E(K_i)$ is the mean number of references to page i, where the page reference following the K_i-th reference to page i is made in accordance with the probability distribution $\{p_1, p_2, ..., p_n\}$, and where $F_m^0(\cdot)$ denotes the value of the miss ratio for these algorithms in the IRM (1).

The validity of (47) follows from two relations. First, we have

$$\frac{1}{T} \sum_{t=1}^{T} \phi(x_t, X_{t-1}^{\text{OPT}}) = \left[\frac{1}{\lambda_T} \sum_{t=1}^{T} \phi(x_t, X_{t-1}^{\text{OPT}}) \right] (\lambda_T/T),$$

where λ_T is the number of transitions in time T in a trajectory of the semi-Markov process (transitions out of state i are performed after K_i sequential references to page i). Second, with probability one

$$\lim_{T \to \infty} \sum_{t=1}^{T} \phi(x_t, X_{t-1}^{\text{OPT}})/\lambda_T = F^0(\text{OPT}) \quad \text{and} \quad \lim_{T \to \infty} \lambda_T/T = (\sum_{i=1}^{n} p_i \nu_i)^{-1}.$$

Formula (48) is proved in the same way as (47).

For the partial Markov model, the above general method may be extended to obtain explicit expressions for $F_m(A)$ under the other replacement algorithms, but the formulas are much more complicated than those for the IRM. For example, if $n = 3$ we get the miss ratio

$$(49) \qquad F_2(\text{FIFO}) = 3/\{(1/p_3)[p_2\nu_2 + (1-p_2)\nu_1] + (1/p_2) \\ [p_1\nu_1 + (1-p_1)\nu_3] + (1/p_1)[p_3\nu_3 + (1-p_3)\nu_2]\},$$

and in the particular case $\nu_1 = \nu_2 = 1$, $\nu_3 = 1/(1-\delta)$, $p_1 = p_2 = p$ and $p_3 = \alpha$ we have the miss ratio

$$F_2(\text{CLIMB}) = \{p^3\alpha + p^4\alpha + p^3\alpha[(1-p)(1-\delta)+\delta] + p^3\alpha^2 + \\ (50) \qquad p\delta[p^2\alpha^2+p^3\alpha] + p^2(1-p)\alpha^2 + [p^2(1-p)\delta/(1-\delta)](\alpha^2+p\alpha)\}/ \\ \{p^3+p^3\alpha+p^2\alpha[(1-p)(1-\delta)+\delta]+p^2\alpha^2 + \delta[p^2\alpha^2+p^3\alpha] + \\ p(1-p)\alpha^2+[p(1-p)\delta/(1-\delta)](\alpha^2+p\alpha)\}.$$

7. Performance Bounds for Replacement Algorithms

Problem Statement. Unless stated otherwise, in this section we assume the IRM of page reference behavior, where the reference probability distribution, $\{p_i\}$ is normally assumed to be unknown, and the pages numbered so that $p_1 \geqslant p_2 \geqslant ... \geqslant p_n$. The basic question to be studied can be stated in general terms as follows: Over all distributions $\{p_i\}$ what is the worst performance of algorithm A relative to OPT? As a by-product of the analysis we are also able to study the reductions in the miss ratio brought about by increases in memory size.

We now make the dependence on the probability distribution explicit in the notation $F_m(A, \{p_i\}) = F_m(A)$. In what follows we examine the ratio $F_m(A, \{p_i\})/F_m(\text{OPT}, \{p_i\})$. Thus, with respect to the first of the above questions, if for some constant γ

$$\sup_{\{p_i\}} F_m(A, \{p_i\})/F_m(\text{OPT}, \{p_i\}) \leqslant \gamma \, ,$$

then A never gives an expected miss ratio higher than $\gamma F_m(\text{OPT}, \{p_i\})$ regardless of the reference probabilities $\{p_i\}$. Results of this type are said to be distribution-free.

General Properties of the Miss Ratio. In our derivation of bounds the following result is useful.

Lemma 3 [FranaW]

(51) $$F_m(\text{OPT}, \{p_i\})/ \sum_{i=m+1}^{n} p_i \leqslant 2 \, .$$

Proof. We use an explicit expression for $F_m(\text{OPT}, \{p_i\})$ to prove the lemma. Writing $\sum_{i=m+1}^{n} p_i = \beta$, we have from (46) that

(52) $$\frac{F_m(\text{OPT}, \{p_i\})}{\sum\limits_{i=m+1}^{n} p_i} \leqslant \frac{1}{\beta} \left[p_m - \frac{p_m^2}{\beta+p_m} \right] + 1$$

It is readily seen that

(53)
$$\frac{\partial}{\partial p_m} (p_m - p_m^2/(\beta + p_m)) \geqslant 0 .$$

Since $p_1 \geqslant p_2 \geqslant ... \geqslant p_n$ we have

(54)
$$p_m \leqslant (1-\beta)/m .$$

The result of the lemma is now implied by (52)-(54). ∎

Within the IRM and a given distribution $\{p_i\}$ we claim that the following two properties are satisfied by any "reasonable" algorithm A:

P1.   Let $p_1' \geqslant p_2' \geqslant ... \geqslant p_n'$, $n \geqslant 3$, be a probability distribution for which $p_i' = p_i$, $1 \leqslant i \leqslant s-1$ and $s+2 \leqslant i \leqslant n$, and $p_s' + p_{s+1}' = p_s + p_{s+1}$. If $F(A, \{p_i'\}) < F(A, \{p_i\})$, then for any $s \geqslant 2$, $F(R(1)A, \{p_i'\}) < F(R(1)A, \{p_i\})$. In words, if $\{p_i'\}$ gives a smaller miss ratio under A, it must also give a smaller miss ratio under the partially preloaded replacement algorithm, $R(1)A$, with page 1 preloaded, i.e. $R_1 = \{1\}$.

P2.   If in P1 we take $s = 1$ and assume that $p_1' > p_1$, then $F(R(1)A, \{p_i'\}) < F(R(1)A, \{p_i\})$. Thus, an increase in p_1 at the expense of p_2 causes the miss ratio for $R(1)A$ to decrease.

It is not hard to verify that OPT and the finite-memory algorithms defined in earlier sections satisfy these properties. Indeed, it is our conjecture that *all* self-organizing finite-memory algorithms satisfy P1 and P2. In any case, in the remainder of this section we restrict ourselves to algorithms satisfying P1 and P2.

As verified below the evaluation of $\sup_{\{p_i\}} F_{m+r}(A, \{p_i\})/ \sum_{i=m+1}^{n} p_i$ can be reduced to finding an upper bound for a function of one variable. Interpreting r as the increase in memory size, we thus obtain a simple measure of the decrease in miss ratios with increases in memory size.

Theorem 14.   Let algorithm A be defined for all $n > m > 0$ and let

(55) $$F_m(R(1)A, \{p_i\}) = (1-p_1)F_{m-1}(A, \{p_2/(1-p_1),..., p_n/(1-p_1)\}) .$$

Then subject to the constraint $\sum_{i=s+1}^{n} p_i = \alpha$, $\max_{\{p_i\}} F_m(A, \{p_i\})$ is achieved for

$$p_1 = p_2 = ... = p_s = (1-\alpha)/s \text{ and } p_{s+1} = p_{s+2} = ... = p_n = \alpha/(n-s).$$

(The proofs of this and subsequent theorems are accumulated at the end of this section.)

Corollary. It follows from Theorem 12 that the result of Theorem 14 is valid for all algorithms being considered. Consequently,

(56) $$\sup_{\{p_i\}} F_{m+r}(A, \{p_i\})/ \sum_{i=m+1}^{n} p_i = \sup_{\{p_i\}} F_{m+r}(A, \{p_i^*\})/[(n-m)\pi] ,$$

where $p_1^* = p_2^* = ... = p_m^* = \theta,$ $p_{m+1}^* = p_{m+2}^* = ... = p_n^* = \pi$ and
$m\theta + (n-m)\pi = 1.$

The following important result follows from (56).

Theorem 15

$$\sup_{\{p_i\}} F_m(A, \{p_i\})/ \sum_{i=m+1}^{n} p_i = \lim_{\epsilon \to 0} \frac{1}{\epsilon} F_m(A, \{\frac{1-\epsilon}{m},..., \frac{1-\epsilon}{m}, \epsilon\}) .$$

Before proceeding we abbreviate our notation and introduce

$$\rho_m(A, \{p_i\}) = F_m(A, \{p_i\})/F_m(\text{OPT}, \{p_i\}) , \quad \rho_m^*(A) = \sup_{\{p_i\}} \rho_m(A, \{p_i\}) ,$$

$$\omega_{m+r}(A, \{p_i\}) = F_{m+r}(A, \{p_i\})/ \sum_{i=m+1}^{n} p_i, \quad \omega_{m+r}^*(A) = \sup_{\{p_i\}} \omega_{m+r}(A, \{p_i\}) ,$$

$$0 \leqslant r < n-m.$$

In the sequel we derive exact expressions for $\rho_m^*(A_\ell^1)$, $\omega_m^*(A_\ell^1)$, $\omega_m^*(A_{m_1}^{m-m_1+1})$, $\omega_{m+r}^*(\text{FIFO})$, and upper bounds for $\omega_{m+r}^*(A_\ell^1)$. Recall that FIFO and LRU correspond, respectively, to the extreme values $\ell = m$ and $\ell = 1$, while the extreme value $m_1 = m$ yields CLIMB.

Theorem 16. For $R_k = \{1, 2,..., k\}$ and $k < m$

$$(57) \qquad \rho_m^* (\text{RA}_\ell^1) = \frac{1}{2} \left[1 + \sum_{i=1}^{m-k-\ell} \frac{1}{i} + \frac{\ell}{m-k-\ell+1} \right]$$

(where $k \leqslant \ell \leqslant m-k$).

In particular, the theorem asserts for $\ell = m-k$ that

$$(58) \qquad \rho_m^* (\text{RFIFO}) = (m-k+1)/2$$

and for $\ell = 1$

$$(59) \qquad \rho_m^* (\text{RLRU}) = \frac{1}{2} \left[1 + \sum_{i=1}^{m-k} \frac{1}{i} \right].$$

From (58) and (59) we observe that the relative difference between FIFO or LRU and an optimal algorithm can be very large. However, the $O(m-k)$ behavior for FIFO is, of course, significantly worse than the $O(\log(m-k))$ behavior for LRU.

Corollary 1. If $R_k = \{1, 2,..., k\}$, $k < m$, then

$$(60) \qquad \omega_{m+r}^* (\text{RA}_\ell^1) \leqslant \frac{1}{2} \left[1 + \sum_{i=1}^{m+r-k-\ell} \frac{1}{i} + \frac{\ell}{m+r-k-\ell+1} \right] \left[1 - \frac{r}{n-m} \right],$$

with equality being achieved when $r = 0$. This inequality is proved by means of (57) as follows:

$$\omega_{m+r}^* (\text{RA}_\ell^1) \leqslant \sup_{\{p_i\}} \left[F_{m+r}(\text{RA}_\ell^1, \{p_i\})/F_{m+r}(\text{OPT}, \{p_i\}) \right] \sup_{\{p_i\}} \left[F_{m+r}(\text{OPT}, \{p_i\})/ \sum_{i=m+1}^{n} p_i \right]$$

from which (60) follows directly. The assertion of equality for $r = 0$ is implied by Theorem 15.

Corollary 2. For the partial Markov and the simplest semi-Markov models

$$\rho_m^* (\text{LRU}) = \frac{1}{2} \left[1 + \sum_{i=1}^{m} \frac{1}{i} \right],$$

and consequently the LRU algorithm becomes no worse relative to an optimal one when the IRM is replaced by the simplest semi-Markov model.

The proof of the corollary follows directly from (47), (48), and (59).

Theorem 17.

(61) $\omega^*_{m+r}(\text{FIFO}) = 1 + [m/(r+1)]$.

This theorem shows that FIFO requires a very considerable increase in memory size to provide a page fault rate close to optimal in a memory of size m. The following two theorems give different upper bounds for $\omega^*_{m+r}(\text{LRU})$.

Theorem 18.

(62) $\omega^*_{m+r}(\text{LRU}) \leqslant [1 + \sum_{i=1}^{m} 1/i][1-r/(n-m)]$.

Theorem 19 [FranaW].

(63) $\omega^*_{m+r}(\text{LRU}) \leqslant 1 + (m/r)[1-(r+1)^{-1}]^{r+1}$.

For purposes of approximation note that $\lim_{r \to \infty} [1-(r+1)^{-1}]^{r+1} = 1/e$.

The bounds of (60), (62) and (63) reveal that even a moderate increase in memory size with the LRU algorithm gives a page fault rate close to optimal for the original memory size. If we let $0 < m \leqslant n-r$, then for $r = 1$ the bound (60) is better than (63) for $\ell = 1$ and for any m and n, as Fig. 13 shows. The bound (60) is also better than (63) when $n-m-r$ is sufficiently small. For example, for $r = 2$ we should have $n-m \leqslant 4$, while for $r = 10$ we should have $n-m \leqslant 7$. In the general case there exists a number m_0 such that (60) is worse than (63) for $m < m_0$ but better for $m \geqslant m_0$. For example, we have $m_0 = 6$ when $n = 100$ and $r = 2$.

The following theorem shows that there exists a family of algorithms within the IRM for which $\rho_m(A)$ does not increase with m, and whose miss ratios do not differ greatly from that of the optimal algorithm for any distribution $\{p_i\}$.

Theorem 20. For $1 \leqslant m_1 < m$

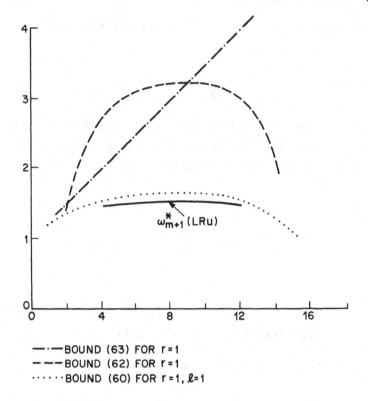

Figure 13 -   Upper bounds for the *LRU* miss ratio with memory size
              $m + 1$.

(64) $$\omega_m^*(A_{m_1}^{m-m_1+1}) = 2 \, .$$

In particular, for $m_1 = 1$ we have the CLIMB algorithm and

(65) $$\omega_m^*(\text{CLIMB}) = 2 \, .$$

By virtue of (64), even a simple A_{m-1}^2 algorithm defined by two priority subsets L_1 and L_2, where $|L_1| = m-1$ and $|L_2| = 1$, gives a miss ratio not far from optimal. It follows from (65) that no increase in complexity of the algorithm due to increases in the number of priority subsets L_i reduces the value of $\omega_m^*(A)$.

Remark [Kog4]. For the partial Markov model, LRU may be optimal within the algorithms A_ℓ^h for certain parameter values. For example, when $m = 2$ the only algorithms to consider are FIFO, LRU and CLIMB. From (49) and (50) we get

$$\lim_{\substack{\alpha \to 0 \\ \delta \to 1}} F_2(\text{CLIMB})/F_2(\text{LRU}) = 7/5 ,$$

$$\lim_{\substack{\alpha \to 0 \\ \delta \to 1}} F_2(\text{FIFO})/F_2(\text{LRU}) = 6/5 .$$

For $m > 2$ we consider the self-organizing algorithm $\tilde{A}$ defined by the parameters $h = 2$, $T_1 = T_2 = \infty$, $\mu_1 = \ell_1 = m-2$, and $D_2 = i_{22}$. It can be shown that $\rho_m^*(\tilde{A})$ is independent of m, whereas $\rho_m^*(\text{LRU})$ increases as $\log m$ by virtue of (59).

Proofs of Theorems.

Proof of Theorem 14. Let $p_i' = p_{i+1}' = (p_i + p_{i+1})/2$ where $p_i \geqslant p_{i+1}$. First, we prove that for any distribution $\{p_i\}$, $n > m$, and $1 \leqslant i \leqslant n-1$

(66) $$F_m(A, \{p_1,..., p_{i-1}, p_i', p_{i+1}', p_{i+2},..., p_n\}) \geqslant F_m(A, \{p_i\}) .$$

For $i = 1$ we note that (66) follows from properties P1 and P2; for $i \geqslant 2$ we prove (66) by induction on m. We use (46) to verify (66) for the basis $m = 1$. Assuming that the inequality holds for $m = 2, 3,..., k$, then from (55) and (66) with $m = k$ it follows that

(67) $$F_{k+1}(R(1)A, \{p_1,..., p_{i-1}, p_i', p_{i+1}', p_{i+2},..., p_n\}) \geqslant F_{k+1}(R(1)A, \{p_i\}) .$$

Hence, from property P1 we find that (66) is true for $m = k+1$.

Now let $\mathbf{p}_{1,1} = \{p_1^{(1)}, p_2^{(1)},..., p_n^{(1)}\}$ denote the initial distribution with $\sum_{j=s+1}^{n} p_j^{(1)} = \alpha$, and let $\mathbf{p}_{1,N}, \mathbf{p}_{2,N},..., \mathbf{p}_{s-1,N}, \mathbf{p}_{s+1,N},..., \mathbf{p}_{n-1,N}$ be a sequence of distributions such that

(68) $$\mathbf{p}_{i,N} = (p_1^{(N)},..., p_{i-1}^{(N)}, \tilde{p}_i^{(N-1)}, \tilde{p}_{i+1}^{(N-1)}, p_{i+2}^{(N-1)},..., p_n^{(N-1)}) ,$$

where $\tilde{p}_i^{(N-1)} = \tilde{p}_{i+1}^{(N-1)} = (p_i^{(N-1)} + p_{i+1}^{(N-1)})/2$ and $p_{i-1}^{(N-1)} = \tilde{p}_{i-1}^{(N-1)}$. Further, let $\mathbf{p}^* = \{p_i^*\}$, where $p_1^* = p_2^* = ... = p_s^* = (1-\alpha)/s$ and $p_{s+1}^* = p_{s+2}^* = ... = p_n^* = \alpha/(n-s)$. Then

(69) $$\lim_{N \to \infty} \mathbf{p}_{n-1,N} = \mathbf{p}^* ,$$

and the statement of the theorem is implied by (66), (68) and (69).

Proof of Theorem 16. We assume for simplicity that $k = 0$, and consider separately the cases $\ell < m$ and $\ell = m$. We first prove (57) for $\ell = 1$. Generalization to the case $\ell < m$ requires certain minor changes in the proof below.

We have to show that

$$
(70) \qquad \rho_m(\text{LRU}) = \frac{1}{2}\left[1 + \sum_{i=1}^{m}\frac{1}{i}\right].
$$

We prove (70) for $n = m+1$; a trivial induction extends the result to arbitrary n. We need the following lemma whose routine proof is left to the interested reader.

Lemma 4. Let $f(x)$ be a convex decreasing function, i.e. $f'(x) < 0$ and $f''(x) > 0$. Then for $0 \leqslant x \leqslant c/2$ we have $f(x)+f(c-x) \leqslant f(0)+f(c)$.

From (35), (45) and (46) we obtain the bound

$$
\rho_m^*(\text{LRU}, \{p_i\}) \leqslant \rho_m^*(\text{LRU}) \sum_{i=1}^{m-1} p_i
$$

$$
(71) \qquad + \frac{p_m F_m(R(m)\text{LRU}, \{p_i\}) + p_{m+1}F_m(R(m+1)\text{LRU}, \{p_i\})}{2p_m p_{m+1}/(p_m+p_{m+1})}, \quad m \geqslant 2
$$

and $\rho_1^*(\text{LRU}) = 1$. Let us put

$$
g_1(p_1,..., p_m, p_{m+1}) = F_m(R(m)\text{LRU}, \{p_i\})/p_{m+1}
$$

and

$$
g_2(p_1,..., p_m, p_{m+1}) = F_m(R(m+1)\text{LRU}, \{p_i\})/p_m .
$$

We apply the lemma to get

$$
g_1(p_1,..., p_m, p_{m+1}) + g_2(p_1,..., p_m, p_{m+1}) \leqslant
$$

$$
g_1(p_1,..., p_{m-1}, p_m+p_{m+1},0) + g_2(p_1,..., p_{m-1}, p_m+p_{m+1}, 0) ,
$$

where for $i = 1, 2$

$$
g_i(p_1,..., p_{m-1}, p_m+p_{m+1}, 0) = \lim_{a \to 0} g_i(p_1,..., p_{m-1}, p_m+p_{m+1}-a, a) .
$$

Then for the set of pages $\{1, 2,..., m\}$ and the probability distribution $Pr\{x_t = m\} = p_m + p_{m+1}$ and $Pr\{x_t = i\} = p_i$, $1 \leqslant i \leqslant m-1$, we have

$$(p_m + p_{m+1}) g_2(p_1,..., p_{m-1}, p_m + p_{m+1}, 0) = F_{m-1, m}(\text{LRU}, \{p_i\})$$

(here it is not assumed that $Pr\{x_t = i+1\} \leqslant Pr\{x_t = i\}$). From (35) and (41) it follows that

(72)
$$F_{m-1, m}(\text{LRU}, \{p_i\}) \leqslant F_{m-1, m}(\text{FIFO}, \{p_i\})$$
$$= m/[1/p_1 +...+1/p_{m-1} + 1/(p_m + p_{m+1})] \leqslant 1/m .$$

Now the expression for $g_1(p_1,..., p_{m-1}, p_m + p_{m+1}, 0)$ can be written as

(73)
$$g_1(p_1,..., p_{m-1}, p_m + p_{m+1}, 0) = 2 \lim_{a \to 0} \frac{F_{m-1, m}(\text{LRU}, \mathbf{p}(a))}{F_{m-1, m}(\text{OPT}, \mathbf{p}(a))} ,$$

where

$$\mathbf{p}(a) = \left\{ \frac{p_1}{1 - p_m - p_{m+1} + a} ,..., \frac{p_{m-1}}{1 - p_m - p_{m+1} + a} , \frac{a}{1 - p_m - p_{m+1} + a} \right\}$$

From (73) we have the bound

(74)
$$g_1(p_1,..., p_{m-1}, p_m + p_{m+1}, 0) \leqslant 2 \rho_{m-1}^*(\text{LRU}) .$$

On substituting bounds (72) and (74) into (71) we obtain

$$\rho_m^*(\text{LRU}) \leqslant \rho_{m-1}^*(\text{LRU}) + 1/2m .$$

Since $\rho_1^*(\text{LRU}) = 1$ we obtain the desired inequality

$$\rho_m^*(\text{LRU}) \leqslant \frac{1}{2} \left[1 + \sum_{i=1}^{m} \frac{1}{i} \right], \quad m \geqslant 1 .$$

Finally, if we let $\mathbf{p}(\epsilon)$ denote the distribution $p_1 = p_2 = ... = p_m = (1-\epsilon)/m$, $p_{m+1} = \epsilon$, $p_{m+2} = ... = p_n = 0$, then it is straightforward to verify that

$$\lim_{\epsilon \to 0} \rho_m(\text{LRU}, \mathbf{p}(\epsilon)) = \frac{1}{2} \left[1 + \sum_{i=1}^{m} \frac{1}{i} \right],$$

so that (70) is proved.

We now prove (57) assuming $\ell = m$ and $k = 0$. From (41) and (46) we have

$$(75) \qquad \frac{F_m(\text{FIFO}, \{p_i\})}{F_m(\text{OPT}, \{p_i\})} = \frac{m+1}{2} \cdot \frac{\displaystyle\sum_{\{i_1,\ldots,\,i_{m+1}\}\in\Lambda'_{m+1,\,n}} p_{i_1}\cdots p_{i_{m+1}} \sum_{i=m}^{n} p_i}{\displaystyle\sum_{\{i_1,\ldots,\,i_m\}\in\Lambda'_{m,\,n}} p_{i_1}\cdots p_{i_m} \sum_{1\leqslant j_1 < j_2 \leqslant m} p_{j_1} p_{j_2}}.$$

We now prove that each term in the numerator of (75) occurs also in the denominator, and therefore $(m+1)/2$ is an upper bound on the right hand side of (75).

The numerator of (75) consists of terms of the two types

$$(76) \qquad p_{i_1} p_{i_2}\cdots p_{i_{m+1}} p_{i_{m+2}}$$

and

$$(77) \qquad p_{i_1} p_{i_2}\cdots p_{i_m} p_{i_s}^2 \,,$$

where $i_k \neq i_\ell$ for $k \neq \ell$ and $i_s = m, m+1,\ldots, n$. The total number of distinct terms of the two types is $\binom{n}{m+2}$. Let α_j, $1 \leqslant j \leqslant m-1$, be the elements of some permutation of $(1, 2,\ldots, m-1)$ and let $0 \leqslant k \leqslant m-1$. Then the product (76) can be represented as

$$(78) \qquad p_{\alpha_1} p_{\alpha_2}\cdots p_{\alpha_k} p_{i_{k+1}} p_{i_{k+2}}\cdots p_{i_{m+1}} p_{i_{m+2}} \,.$$

This occurs in the numerator of (75) $m-k+2$ times and can be seen in the denominator in the following form:

$$(79) \qquad p_{\alpha_1} p_{\alpha_2}\cdots p_{\alpha_k} p_{i_{k+1}} p_{i_{k+2}}\cdots p_{i_m} (p_{i_{m+1}} p_{i_{m+2}}) \,, \quad 0 \leqslant k \leqslant m-1 \,.$$

Since $i_\ell > m$, $\ell = k+1,\ldots, m+2$, in (78) and (79), we can identify all distinct pairs $p_i p_j$ within the parentheses, where $i, j = i_{k+1}, i_{k+2},\ldots, i_{m+2}$. Consequently, there are $\binom{m-k+2}{2}$ identical products of the form (78) in the denominator of (75). Since for $0 \leqslant k \leqslant m-1$ the inequality $\binom{m-k+2}{2} \geqslant m-k+2$ holds, each term of type (76) in the numerator of (75) occurs in the denominator.

We now show that each product of the form (77) occurs in the denominator of (75). By definition of the set $\Lambda'_{m+1,\,n}$ there are no two identical products of the form (77) in the numerator of (75). Each product (77) consists of $m+1$ distinct factors and these include two cofactors p_{i_s} and p_{i_ℓ} for which $i_s, i_\ell \geqslant m$. Without loss of generality we can assume that $p_{i_\ell} = p_{i_m}$. In the outer sum of the denominator in (75) we have a term of the form $p_{i_1}p_{i_2}\cdots p_{i_{m-1}}p_{i_s}$, and by virtue of $i_s \geqslant m$, $i_m \geqslant m$, $i_s \neq i_m$ this is multiplied by $p_{i_s}p_{i_m}$ from the inner sum of the denominator. Comparing with (77) we see that each product of the form (77) is present in the denominator of (75). Thus,

$$(80) \qquad \frac{F_m(\text{FIFO}, \{p_i\})}{F_m(\text{OPT}, \{p_i\})} \leqslant (m+1)/2 .$$

Finally, it is readily seen that

$$\lim_{\epsilon \to 0} \rho_m(\text{FIFO}, \mathbf{p}(\epsilon)) = (m+1)/2 ,$$

which together with (80) yields (57) for $\ell = m$ and $k = 0$.

Proof of Theorem 17. We get directly from (41)

$$\frac{F_{m+r}(\text{FIFO}, \{p_i\})}{\displaystyle\sum_{i=m+1}^{n} p_i} \leqslant \frac{m+r+1}{r+1} = 1 + \frac{m}{r+1} .$$

Following [FranaW] we consider the distribution $\mathbf{p}(\beta)$ defined by

$$p_i = \begin{cases} (1-\beta)/m , & 1 \leqslant i \leqslant m , \\ \beta/(n-m) , & m < i \leqslant n , \end{cases}$$

where $n \geqslant m+r$ and $0 < \beta \ll 1$. Using (41) we obtain

$$\lim_{\substack{\beta \to 0 \\ n \to \infty}} \rho_{m+r}(\text{FIFO}, \mathbf{p}(\beta)) = 1 + m/(r+1) ,$$

by which the theorem is proved.

Proof of Theorem 18. From (56) it follows that

$$\rho_{m+r}^*(\text{LRU}) = \sup_{\{p_i\}} \rho_{m+r}(\text{LRU}, \{p_i^*\}) ,$$

where $p_i^* = \theta$, $1 \leqslant i \leqslant m$, $p_i^* = \pi$, $m < i \leqslant n$ and $m\theta + (n-m)\pi = 1$.

Therefore, to prove the theorem it is sufficient to show that

$$(81) \qquad \sup_{\{p_i\}} \rho_{m+r}(\text{LRU}, \{p_i^*\}) \leqslant [1 + \sum_{i=1}^{m} 1/i][1 - r/(n-m)] .$$

Let Ψ_{r+s} be a subset of $\Lambda_{r+s+1,\,n}$ satisfying the following conditions. The permutation $(i_1, i_2, \ldots, i_{r+s+1})$ belongs to Ψ_{r+s} if and only if $m+1 \leqslant i_{r+s+1} \leqslant n$ and there exists a nonnegative integer $j \leqslant s$ such that $(i_1, i_2, \ldots, i_{r+s})$ consists of j numbers from the set $(1,2,\ldots,m)$ and $r+s-j$ numbers from the set $(m+1, m+2, \ldots, n)$. Furthermore, for each element $(i_1, i_2, \ldots, i_{r+s+1}) \in \Psi_{r+s}$ let $\Omega_{m-s} = \Omega_{m-s}(i_1, i_2, \ldots, i_{r+s})$, $m-s \geqslant 1$, be the set of all permutations of $m-s$ distinct numbers selected from the set $(1,2,\ldots,m) \setminus (i_1, i_2, \ldots, i_{r+s})$, with $\Omega_0 = \varnothing$. We put

$$(82) \qquad \eta_s = \sum_{\Psi_{r+s}} \frac{p_{i_1}^* \cdots p_{i_{r+s}}^* \, p_{i_{r+s+1}}^*}{(1-p_{i_1}^*) \cdots (1-p_{i_1}^* - \ldots - p_{i_{r+s}}^*)} \times$$

$$\sum_{\Omega_{m-s}} \frac{p_{i_{r+s+2}}^* \cdots p_{i_{m+r+1}}^*}{(1-p_{i_1}^* - \ldots - p_{i_{r+s+1}}^*) \cdots (1-p_{i_1}^* - \ldots - p_{i_{m+r-1}}^*)} \, , \, s=0,1,\ldots,m \, .$$

Then from (35) and (82) we get

$$(83) \qquad\qquad F_{m+r} \, (\text{LRU}, \{p_i^*\}) = \sum_{s=0}^{m} \eta_s \, .$$

Comparing (36) and (82) we have

$$(84) \qquad \frac{\eta_m}{(n-m)\pi} \leqslant \frac{n-m-r}{n-m}, \qquad \frac{\eta_{m-1}}{(n-m)\pi} \leqslant \frac{n-m-r}{n-m} \, .$$

From the inequality $(1 - p_{i_1}^*, -\ldots-p_{i_{r+s}}^*) \geqslant (m-s)\theta$ we get an upper bound for $\eta_s, \, 0 \leqslant s \leqslant m - 2$:

$$\eta_s \leqslant \frac{(n-m-r)\pi}{m-s} \sum_{\Psi_{r+s}} \frac{p_{i_1}^* \cdots p_{i_{r+s}}^*}{(1-p_{i_1}^*) \cdots (1-p_{i_1}^* -\ldots-p_{i_{r+s-1}}^*)}$$

$$(85) \qquad \times \sum_{\Omega_{m-s}} \frac{p_{i_{r+s+2}}^* \cdots p_{i_{m+r}}^*}{(1-p_{i_1}^* -\ldots-p_{i_{r+s+1}}^*) \cdots (1-p_{i_1}^* -\ldots-p_{i_{m+r-1}}^*)} \, .$$

Using the properties of $\{p_i^*\}$ and comparing the terms in the inner sum on the right-hand side of (85) with the components of the vector q^{RLRU}, where $R_k = R_{r+s+1} = (i_1, i_2, \ldots, i_{r+s+1})$, we have

$$\sum_{\Omega_{m-s}} \frac{p_{i_{r+s+1}}^* \cdots p_{i_{m+r}}^*}{(1-p_{i_1}^* -\ldots-p_{i_{r+s+1}}^*) \cdots (1-p_{i_1}^* -\ldots-p_{i_{m+r-1}}^*)} \leqslant 1$$

Then from a comparison of the right-hand sides of (85) and (36) we get for $0 \leqslant s \leqslant m-2$ the inequality

$$\eta_s/[(n-m)\pi] \leqslant (n-m-r)/[(m-s)(n-m)]$$

from which (81) follows in view of (83) and (84).

Proof of Theorem 19. Page i is absent from a memory of size $m + r$ when $r + s$ ($s \geqslant 1$) distinct pages with numbers $j > m$ and $m - s$ distinct pages $\ell, \ell \leqslant m$, have been referenced since the most recent reference to page i. Thus a necessary condition for the absence of page i from memory is that at least $r + 1$ distinct pages $j, j > m$, have been referenced since the last reference to i. For $\sum\limits_{i=m+1}^{n} p_i = \alpha$ the probability of this condition being satisfied does not exceed

$$g(p_i) = [\alpha/(\alpha+p_i)]^{r+1},$$

since the probability of a reference to any of the pages $j > m$ since the last reference to page i is $\alpha/(\alpha + p_i)$. Next, we use the fact assured by the ergodic theorem that the LRU miss ratio $F(LRU)$ can be represented as

$$F(LRU) = \sum_{i=1}^{n} p_i \beta_i$$

where $\beta_i = \lim\limits_{t \to \infty} Pr\{i \in X_{t-1}^{LRU}\}$.

It follows from the definition of $g(p_i)$ that $\beta_i \leqslant g(p_i)$. But

$$\max_{0 \leqslant p_i \leqslant 1} p_i g(p_i) = (\alpha/r)[1 - (r+1)^{-1}]^{r+1}.$$

Therefore,

$$F_{m+r} (LRU)/ \sum_{i=m+1}^{n} p_i \leqslant 1 + (\sum_{i=1}^{m} p_i \beta_i/\alpha)$$
$$\leqslant 1 + (m/r)[1 - (r+1)^{-1}]^{r+1}.$$

Proof of Theorem 20. For simplicity we prove the statement of (64) for $m_1 = 1$, i.e. for the CLIMB algorithm. There is no essential difficulty in generalizing to the case $1 \leqslant m_1 < m$. From property $P1$ it follows that

$$\rho_m^*(CLIMB) = \sup_{\{\bar{p}_i\}} \rho_m (CLIMB, \{\bar{p}_i\})$$

where $\{\tilde{p}_i\}$ is a family of distributions $(\tilde{p}_1, \tilde{p}_2, \ldots, \tilde{p}_n)$ such that

$$\tilde{p}_{m+1} = \tilde{p}_{m+2} = \cdots = \tilde{p}_n = \pi .$$

To prove

(86) $$\sup_{\{\tilde{p}_i\}} \rho_m \, (\text{CLIMB}, \{\tilde{p}_i\}) = 2$$

we use the inequality

(87) $$a^k b \leqslant \frac{k}{k+1} a^{k+1} + \frac{1}{k+1} b^{k+1} ,$$

where k is an integer and $a, b > 0$. This inequality is readily proved by induction. From (42) with $m_1 = 1$ and $\{p_i\} = \{\tilde{p}_i\}$ we have

(88) $$I_m \, (\text{CLIMB}) = \frac{\dfrac{1}{\pi} \displaystyle\sum_{(i_1, \ldots, i_m) \in \Delta_{m,n}} \tilde{p}_{i_1}^m \tilde{p}_{i_2}^{m-1} \cdots \tilde{p}_{i_m} \sum_{s=m+1}^{n} \tilde{p}_{i_s}}{(n-m) \displaystyle\sum_{(i_1, \ldots, i_m) \in \Delta_{m,n}} \tilde{p}_{i_1}^m \tilde{p}_{i_2}^{m-1} \cdots \tilde{p}_{i_m}}$$

$$\equiv \frac{W}{(n-m) V}$$

Consider the sum in the numerator of (88). Denote by $d_{m-\ell}$ the first $m-\ell$ cofactors in a term of the sum whose successive $\ell+1$ cofactors take the form

$$\tilde{p}_{i_k}^\ell \tilde{p}_{i_{k+1}}^{\ell-1} \cdots \tilde{p}_{i_{k+\ell-1}} \tilde{p}_{i_{k+\ell}} = \pi h_\ell, \ \ell \geqslant 1,$$

where $m+1 \leqslant i_k \leqslant n$ and $1 \leqslant i_{k+j} \leqslant m, \ j \geqslant 1$. Applying (87) we have

$$\tilde{p}_{i_k}^\ell \tilde{p}_{i_{k+1}}^{\ell-1} \cdots \tilde{p}_{i_{k+\ell-1}} \tilde{p}_{i_{k+\ell}} = \pi \tilde{p}_{i_{k+1}}^{\ell+1} \tilde{p}_{i_{k+\ell}} \pi^{\ell-1} \tilde{p}_{i_{k+2}}^{\ell-2} \cdots \tilde{p}_{i_{k+\ell-1}}$$

$$\leqslant \pi \left[\frac{\ell-1}{\ell} \tilde{p}_{i_{k+1}}^\ell \tilde{p}_{i_k}^{\ell-1} \tilde{p}_{i_{k+2}}^{\ell-2} \cdots \tilde{p}_{i_{k+\ell-1}} + \frac{1}{\ell} \tilde{p}_{i_{k+\ell}}^\ell \tilde{p}_{i_k}^{\ell-1} \tilde{p}_{i_{k+2}}^{\ell-2} \cdots \tilde{p}_{i_{k+\ell-1}} \right]$$

$$= \pi \left[\frac{\ell-1}{\ell} \gamma_{1\ell} + \frac{1}{\ell} \gamma_{2\ell} \right] = \pi \gamma_\ell .$$

Now denote by d the expression derived from W by substituting $d_{m-\ell} \gamma_\ell$ for $d_{m-\ell} h_\ell$ ($\ell \geqslant 2$). From (43) with $m_1 = 1$ and $\{p_i\} = \{\tilde{p}_i\}$ we see that $d_{m-\ell} \gamma_{i\ell}$ ($i=1,2$) occurs not more than $2(n-m)$ times in d and once in V. In addition, each term $d_{m-1} \tilde{p}_{i_{k+\ell}}$, $i_{k+\ell} \leqslant m$, occurs not more than $n-m$ times in W

and once in V. The same holds for d_m. Therefore,

$$(89) \qquad\qquad \sup_{\{\tilde{p}_i\}} \rho_m \, (\mathrm{CLIMB}, \, \{\tilde{p}_i\}) \leqslant 2 \, .$$

Moreover, it is readily seen that

$$(90) \qquad\qquad \lim_{\epsilon \to 0} \rho_m \, (\mathrm{CLIMB}, \, \mathbf{p}(\epsilon)) = 2 \, .$$

Equation (86) follows from (89) and (90).

8. Additional Topics

The majority of the analytical results of this section have been obtained in the framework of the independent reference model. Other interesting applications of the IRM may be found in [FagE, Rao]. The main result of [FagE] is a theorem proving the independence of the expected working-set miss ratio and page size. In [Rao] the performances of different cache memory organizations (Fully Associative, Direct Mapping, Set Associative, and Sector) are analyzed. First it is shown how the fault rates for these caches can be calculated. It is then shown how the expressions can be simplified to obtain distribution-free results on cache performance. Such bounds are derived for the Direct Mapping cache and the Set Associative LRU cache. For the Sector cache this bound was derived in terms of the upper bound on its equivalent Fully Associative cache performance. The bounds for the Direct Mapping cache were used to compute a close upper bound on the amount of cache memory needed to guarantee a given level of performance.

Certain other models, in particular the LRU stack model and its generalizations, have been applied to the evaluation and comparison of replacement algorithms under different workloads [ChuO, ChusH, Eas2, FerLW, GupF, SheS]. The latter, simple model of program behavior has a number of useful features [CofD, Spi]. In this model, the set of pages is organized into a stack in LRU order. Page reference behavior is then represented by a sequence of stack "distances", i.e. index positions in the stack [MatGST]. This sequence is assumed to consist of independent, identically distributed random variables. While the model provides for locality of reference, it fails to represent changes in the size of a locality or abrupt changes in the content of a locality. In [Smi2] algorithms are introduced for calculating the expected number of faults when a program's paging behavior is described by either the LRU stack model or the IRM, and the replacement algorithm is MIN or the algorithm VMIN described below.

The VMIN algorithm [PriF] is an algorithm (unrealizable in real time) which minimizes both the average memory use and the number of page faults over a given virtual (process) time interval. It works as follows: For a fixed T, all resident pages that are not to be referenced in the following T time units are removed. New pages are brought into main memory only on demand. By varying T, a piecewise linear curve is traced out in the plane of page faults vs. average memory size. For the IRM it was shown that the average memory size for parameter T when using the VMIN algorithm is

$$n - \sum_{i=1}^{n} (1 - p_i)^T (1 + Tp_i),$$

which, we observe, is less than for WS, as given in (10) with $\delta = 0$. The page fault rate is given by (13). In other cases the algorithms are similar to those obtained by the general method of section 6. These results complement those of section 6, although the complexity of formulas in [Smi2] makes direct calculation for most programs difficult or impractical.

An unrealizable algorithm similar in purpose to VMIN is developed in [BudDMS]. This algorithm, DMIN, has the following properties. A dynamic (time-varying) size of allocation is computed by DMIN in order to minimize the space-time product of main memory allocated to a program during execution. The algorithm is a function of one parameter, the reactivation time, which is the average time from the occurrence of a page fault to the restart of execution of the corresponding program. DMIN is used primarily as a performance measurement tool for bounding the performance of dynamic algorithms such as the WS algorithm and the page-fault-frequency (PFF) algorithm. (With PFF the locality set of pages is estimated by observing the page fault rate [ChuO].)

In contrast to previous studies of program behavior, a general stochastic jump process is exploited in [TzeG] to describe the page reference generator. First a model of memory management is formally defined as three component processes: the program behavior, memory allocation, and control processes. Then relations linking the evolution of the memory allocation process with the other two processes are derived. Necessary and sufficient conditions for an optimal control policy are given as a set of (Bellman) optimality equations, which can be solved numerically for small problem sizes. An analytic solution is presented for the case of stationary ranking pages. This result complements those in section 5.

In [VolMHKR] a new technique is proposed for modeling program behavior. The model implies a definition of reference locality that does not require the prior specification of a fixed-size time window to compute working sets and cache miss ratios.

One important topic that was excluded from this chapter is the problem of reorganizing programs to improve observed locality. (See [HatG, AbuKL], for example. Other key papers are briefly reviewed in [Den3].) While there are several papers [BarS, Eas1, ParF, Rao, YueW] using stochastic models of program reorganization, and in particular the independent [YueW, Rao] and Markov [Eas1, BarS] reference models, the results are based for the most part on continuous or discrete optimization techniques rather than on stochastic analysis.

The self-organizing replacement algorithms presented in section 2 were first published in the period 1975-78 [AveBK1, AveBK2, Kog3, Kog4], and they aroused the interest of probability theorists [Die, Let]. In a recent paper [BabF] a subfamily of these algorithms, including the algorithms A_ℓ^1, was reconsidered and analyzed in detail. The analysis included trace-driven simulations and a study of implementation questions. Self-organizing algorithms appear to be especially appropriate for the two-level memory management of computer operating systems, and especially for the control of the disk cache [GurK, Kog5, Smi5].

Chapter 5

Paged Two-Level Memories:
Approximate Models and Efficient Computations

1. Preliminaries

In section 4.6 we developed a general method of evaluating the miss ratio for the independent reference model (IRM) of program behavior. This model ignored the fact that there are two essentially different types of program behavior. One applies when the program is referencing within a locality and the other when the program is changing locality. The IRM had one further shortcoming. Even for moderate memory sizes and numbers of program pages, the IRM requirements in computer time and space exceed the available resources when formula (4.34) or its explicit variants, e.g. (4.35) or (4.36), are used. A similar shortcoming obtains in general in the Markov model of program behavior.

Here we pursue two objectives. First, we construct a more realistic and tractable model of program behavior. Second, within this model and those of the preceding chapter we present efficient techniques for approximating the miss ratios of the commonly used replacement algorithms, FIFO, LRU and WS.

In the next section we propose a nearly decomposable model of program behavior which attempts to keep the advantages of the IRM and simple Markov models in describing program behavior within localities. The model also includes a mechanism that takes into account transitions among different localities, an aspect of program behavior which is now known to be as important as the concept of locality itself [Den3]. Our model differs from the nearly completely decomposable model proposed by Courtois [Cou,CouV] in one key respect. The latter assumes that programs consist only of localities and that each page belongs to exactly one locality, whereas in our model we have page sets that do not belong to any locality. Such sets are referenced when the program changes locality.

In what follows we work out an extension of the Simon-Ando approximations [Cou] for the stationary probabilities of the nearly decomposable Markov chain.

156

These approximate probabilities are then used in deriving approximations for miss ratios and working-set size distributions. We show that these performance measures can be expressed as weighted sums of the respective measures for the localities. The results we present were first published in [AveGK, Kog6, KogT]. We also describe recent results [Fag, FagP, Tze] on the efficient calculation of miss ratios for FIFO and LRU replacement algorithms, when program behavior within a locality is modeled by the IRM and Markov models.

2. A Nearly Decomposable Model of Program Behavior

We analyze first the case of strictly disjoint localities. When representing a program with localities, a Markov matrix such as the matrix $P = ||p_{ij}||_{i,j=1}^{n}$ introduced in section 4.3 enjoys a special property that so far has not been exploited. Suppose that its rows and columns are arranged so that pages of the same locality have consecutive entries. Then by definition, if the subset of pages $U \subset X$ is a locality, then all elements p_{ij}, $i,j \in U$, will be relatively large and all elements p_{ij}, $i \in U$ and $j \in X \backslash U$, will be comparatively small, since they are probabilities of referencing while in U a page that does not belong to U. More precisely we describe the program behavior by a matrix $P^\epsilon = ||p_{ij}^\epsilon||_{i,j=1}^{n}$ admitting the representation

$$(1) \qquad\qquad P^\epsilon = P^o + \epsilon\Phi,$$

where P^ϵ is the transition matrix of a regular Markov chain[†]; $P^o = ||p_{ij}^o||_{i,j=1}^{n}$ is a stochastic matrix whose elements are independent of ϵ; the elements of the matrix Φ are also independent of ϵ and do not exceed unity in absolute value; and ϵ is small compared to the elements of P^o (ϵ bounds the probability of leaving a locality). It is assumed that the structure of the matrix P^ϵ is such that as $\epsilon \to 0$ the limiting matrix P^o defines a Markov chain with several ergodic and transient sets of states. The transition matrices P^ϵ and their corresponding Markov chains are called nearly decomposable.

We denote by $U_1, U_2, \ldots, U_K$ the ergodic sets of the chain with transition matrix P^o and by H the set of all its transient states. We combine in the set u_1

---

† A chain is regular when any state can be reached in a finite number of steps from any other state with positive probability [KemS].

those states of H from which, in the chain with transition matrix P^ϵ, the set U_1 can be reached with positive probability. The transient set u_i, $i > 1$, corresponding to U_i is formed in the same way except that H is replaced by $H \backslash \bigcup_{j<i} u_j$. We denote by n_A, n_B the numbers of elements in the sets A, $B \subset X$ and by $Q(A \times B)$ the $n_A \times n_B$ submatrix of $Q = Q(X \times X)$ consisting of those elements at the intersections of rows with numbers from the set A and columns with numbers from the set B. Henceforth, the matrix P^ϵ can be more conveniently expressed in the form

(2)

$$
P^\epsilon =
\begin{bmatrix}
P^o(U_1 \times U_1) & 0 & 0 & 0 & & 0 & 0 \\
P^o(u_1 \times U_1) & P^o(u_1 \times u_1) & P^o(u_1 \times U_2) & P^o(u_1 \times u_2) & \cdots & P^o(u_1 \times U_K) & P^o(u_1 \times u_K) \\
0 & 0 & P^o(U_2 \times U_2) & 0 & \cdots & 0 & 0 \\
0 & 0 & P^o(u_2 \times U_2) & P^o(u_2 \times u_2) & \cdots & P^o(u_2 \times U_K) & P^o(u_2 \times u_K) \\
\vdots & \vdots & \cdots & & \cdots & \vdots & \\
0 & 0 & \cdots & & & P^o(U_K \times U_K) & 0 \\
0 & 0 & & & & P^o(u_K \times U_K) & P^o(u_K \times u_K)
\end{bmatrix}
+ \epsilon R
$$

where the matrices $P^o(U_1 \times U_1), P^o(U_2 \times U_2), \ldots, P^o(U_K \times U_K)$ define regular Markov chains, the non-zero elements of R are confined to the submatrices $R(U_1 \times X), R(U_2 \times X), \ldots, R(U_K \times X)$, and 0 denotes the matrix of zeros. Matrices $P^o(U_1 \times U_1), P^o(U_2 \times U_2), \ldots, P^o(U_K \times U_K)$ are separate models of program behavior within the respective, strictly disjoint localities $U_1, U_2, \ldots, U_K$ while matrices $P^o(u_1 \times U_1), P^o(u_1 \times u_1), P^o(u_1 \times U_2), \ldots, P^o(u_1 \times u_K)$, $P^o(u_2 \times U_2), \ldots, P^o(u_K \times U_K), P^o(u_K \times u_K)$ model behavior outside the localities. It should be noted that the IRM is an acceptable model of program behavior within a locality U whose matrix $P^o(U \times U)$ has little variation in each column. Such localities are likely to exist, since matrices $P^o(U_i \times U_i)$, $i = 1, 2, \ldots, K$, have by definition elements of approximately the same order of magnitude [Cou]. With its analytical advantages the IRM is clearly desirable under these circumstances.

We now treat the case of overlapping localities. Following [Cou,CouV] we consider an example model described by a transition matrix P of the type shown in Fig. 1. It can be assumed that in this program there are three localities $U_1 = (1,2,3,4,5)$, $U_2 = (4,5,6,7)$ and $U_3 = (7,8,9)$, which overlap pairwise. However, this model possesses obvious drawbacks. If $\epsilon \rightarrow 0$, it is natural to assume that the matrix P approaches a stochastic matrix P^o in which the elements represented by o's in Fig. 1

$$
P \;=\;
\begin{array}{c}
\begin{array}{ccccccccc}
1 & 2 & 3 & 4 & 5 & 6 & 7 & 8 & 9
\end{array}\\[4pt]
\left[
\begin{array}{ccccccccc}
x & x & x & x & x & o & o & o & o\\
x & x & x & x & x & o & o & o & o\\
x & x & x & x & x & o & o & o & o\\
x & x & x & x & x & x & x & o & o\\
x & x & x & x & x & x & x & o & o\\
o & o & o & x & x & x & x & o & o\\
o & o & o & x & x & x & x & x & x\\
o & o & o & o & o & o & x & x & x\\
o & o & o & o & o & o & x & x & x
\end{array}
\right]
\end{array}
$$

Figure 1   x is an element of magnitude $O(1)$,

o is an element of magnitude $O(\epsilon)$

will be zero. The Markov chain with the transition matrix P^o has a single ergodic set, coinciding with the whole set of states of the initial chain, because from pages belonging to locality intersections, references to both localities may occur with probabilities of the same order of magnitude. Hence the matrix P^o does not reduce the dimensionality of the problem of finding the stationary distribution. Another, more important drawback of the model is the fact that there are no subsets of pages in which references are concentrated over fairly long time intervals. We will now construct a model with overlapping localities which is free from these drawbacks.

Suppose a program consists of N *modules*, each exhibiting a different referencing behavior, and described as follows. If X_k is the page-set of the k-th module with n_k pages, then $\bigcup_{k=1}^{n} X_k = X$, but in general the intersections of page sets from different modules are non-empty. It is assumed that localities of the k-th module $(k = 1, 2, \ldots, N)$ form noninteresecting sets of pages $U_1(k), \ldots, U_{\eta(k)}(k) \subset X_k$, i.e. $U_i(k) \cap U_j(k) = \varnothing$ when $i \neq j$ and $i, j = 1, 2, \ldots, \eta(k)$. However, there can be pages in a module not belonging to any locality of that module.

The behavior within modules is described by absorbing Markov chains with the following properties:

1) Absorbing states of the k-th chain $(k = 1, 2, \ldots, N)$ are just those elements of $X_k \setminus \sum_{i=1}^{\eta(k)} U_i(k)$ which belong to some module $i \neq k$ (i.e. it is assumed that referencing of the k-th module is completed when a reference occurs to a page of another module which does not belong to any locality of the k-th module);.

2) From each nonabsorbing state of the k-th chain any other state can be reached in a finite number of steps which does not depend on the initial state;

3) The transition matrix $P_k = ||p_{ij}(k)||$ of the k-th chain has the form

$$P_k = P_k^0 + \epsilon \Phi_k,$$

where P_k^0 is a stochastic matrix with elements that are independent of ϵ, the elements of the matrices Φ_k do not exceed unity in absolute value, and ϵ is small by comparison to the elements of P_k^0 and Φ_k;

4) The matrix P_k^0 defines a Markov chain whose ergodic sets, with the exception of the absorbing states, constitute the localities $U_1(k), \ldots, U_{\eta(k)}(k)$.

The last two properties reflect the appearance of localities within modules.

In the example considered above (see Fig. 1) we can distinguish three modules each with one locality: $U_1(1) = (1,2,3,4,5)$, $U_1(2) = (4,5,6,7)$, and $U_1(3) = (7,8,9)$. The page sets of all modules are identical and coincide with the page set X of the program. For the first chain the states 6, 7, 8, and 9 are absorbing, for the second the states 1,2,3,8, and 9 are absorbing, and for the third the states 1, 2, 3, 4, 5, and 6 are absorbing.

Transitions from one module to another can be modeled in several ways. In order to preserve the property of program locality as a whole, we assume that the k-th module is changed only at absorbing states of the k-th chain. The module change mechanism is described by probability distributions

$$\alpha_k(j) = \left[\alpha_{k1}(j), \ldots, \alpha_{kN}(j)\right]$$

given for each absorbing state j of the k-th chain $(k = 1,2,\ldots,N)$. The distributions $\alpha_k(j)$ have the following properties:

(i) $\alpha_{k\ell}(j) \geqslant 0,\ k,\ell = 1,2,\ldots,N$,

(ii) if state j is absorbing for the ℓ-th chain then $\alpha_{k\ell}(j) = 0$,

(iii) $\displaystyle\sum_{\ell=1}^{N} \alpha_{k\ell}(j) = 1$, and

(iv) starting from any set X_k any other set X_ℓ, $\ell \neq k$, can be reached after a finite number of steps which does not depend on X_k.

For the example in Fig. 1 we put

$$\alpha_1(j) = \begin{cases} (0,1,0) & \text{for } j = 6,7 \\ (0,0,1) & \text{for } j = 8,9 \end{cases}$$

$$\alpha_2(j) = \begin{cases} (1,0,0) & \text{for } j = 1,2,3 \\ (0,0,1) & \text{for } j = 8,9 \end{cases}$$

$$\alpha_3(j) = \begin{cases} (0,1,0) & \text{for } j = 4,5,6 \\ (1,0,0) & \text{for } j = 1,2,3 . \end{cases}$$

Returning to the general case we note that the page reference process is not Markovian, since a page can belong to more than one module. On the other hand, with the above assumptions on the module behavior and the module change mechanism, the process $\{(x_t,y_t)\}$ will be a Markov chain, where y_t is the index of the module containing page x_t at time t.

The state space of the Markov chain consists of all possible pairs (i,k), where $i \in X$, and k is the index of the module containing page i. The transition probabilities $p((i,k), (j,\ell))$ of this chain are given by the following equations:

$$(3) \qquad p\left((i,k),(j,\ell)\right) = \begin{cases} p_{ij}(k) & \text{if } \ell = k \text{ and } p_{jj}(k) < 1, \\ 0 & \text{if } \ell \neq k \text{ and } p_{jj}(k) < 1, \\ p_{ij}(k)\alpha_{k\ell}(j) & \text{if } p_{jj}(k) = 1. \end{cases}$$

It follows from the properties of the Markov chains describing module behavior and from the distributions $\alpha_k(j)$ that the chain with transition probabilities (3) is regular and that its transition matrix admits the representation (1). Thus we have obtained a nearly decomposable Markov model of program behavior with overlapping localities.

3. Approximation of the Stationary Probability Vector

To obtain approximations of performance measures in the nearly decomposable model of program behavior we first derive the approximation for the stationary probability vector of any nearly decomposable Markov chain. Suppose the row vector $p^\epsilon = (p_1^\epsilon, \ldots, p_n^\epsilon)$ is the stationary distribution of a chain with transition matrix P^ϵ of form (2). We are interested in the limiting stationary probabilities

$$p_i^* = \lim_{\epsilon \to 0} p_i^\epsilon.$$

The limiting probability p_i^* of referencing page i of locality U can be expressed as a product of
(a) the limiting stationary probability of referencing locality U, which can be found from an aggregated Markov chain, and
(b) the stationary probability of referencing page i in a program whose behavior is described by the transition matrix $P^o(U \times U)$. (The latter approximates the steady-state conditional probability that page i is referenced, given that the reference is directed to locality U).

We begin by constructing the transition matrix Q^ϵ of the aggregated Markov chain, whose states correspond to the sets $U_1, U_2, \ldots, U_K$ and are denoted by the same symbols. Let U be an arbitrary ergodic set of the chain with transition matrix P^o, and let $p^o(U) = (p_1^o(U), \ldots, p_{n_U}^o(U))$ be the stationary probability vector of the matrix $P^o(U \times U)$. Then the elements $q_{U,Y}^\epsilon, Y = U_1, U_2, \ldots, U_K$ of matrix Q^ϵ are calculated by averaging the transition probabilities p_{ij}^ϵ over the distribution $p^o(U)$ and subsequently summing over the set Y:

(4)
$$q_{U,Y}^{\epsilon} = p^o(U)P^{\epsilon}(U \times Y)\mathbf{1}'(n_Y)$$

$$= \begin{cases} 1 + \epsilon p^o(U)R(U \times U)\mathbf{1}'(n_U), & Y = U \\ \epsilon p^o(U)R(U \times Y)\mathbf{1}'(n_Y), & Y \neq U \end{cases}$$

where $\mathbf{1}(n_Y)$ is the row vector of n_Y 1's. Similarly, the elements $q_{U,i}^{\epsilon}$, $i \in y$, $y = u_1, u_2, \ldots, u_K$, are the components of the vector

(5)
$$q_U^{\epsilon}(y) = p^o(U)P^{\epsilon}(U \times y) = \epsilon p^o(U)R(U \times y) .$$

To each submatrix of the form $P^{\epsilon}(y \times Y) = P^o(y \times Y)$, $y = u_1, u_2, \ldots, u_K$, $Y = U_1 U_2, \ldots, U_K$ in Q^{ϵ}, there corresponds one column equal to the sum of the columns of the submatrix. Finally, the submatrices of the form $P^{\epsilon}(y \times z) = P^o(y \times z)$, $y = u_1, u_2, \ldots, u_K$, $z = u_1, u_2, \ldots, u_K$, occur in Q^{ϵ} without change. If L is the total number of states in the locality sets $U_1, U_2, \ldots, U_K$ then the order of matrix Q^{ϵ} is $n - L + K$.

The regularity of the chain with transition matrix Q^{ϵ} readily follows from the regularity of the initial chain with transition matrix P^{ϵ} and the construction of Q^{ϵ}. The components of the vector q^{ϵ} of stationary probabilities of the matrix Q^{ϵ}, corresponding to the sets U_i, $i = 1, \ldots, K$, will be denoted by $q^{\epsilon}(U_i)$, $i = 1, \ldots, K$; and the components of this vector corresponding to the sets u_j, $j = 1, \ldots, K$, will be denoted by $\{q_i^{\epsilon}(u_j), i \in u_j\}$, $j = 1, \ldots, K$.

Theorem 1. For the nearly decomposable Markov chain, $i \in U$ and $U = U_1, U_2, \ldots, U_K$ we have

(6)
$$\lim_{\epsilon \to 0} \frac{p_i^{\epsilon}}{p_i^o(U)q^{\epsilon}(U)} = 1$$

and for $i \in u$, $u = u_1, u_2, \ldots, u_K$, we have

(7)
$$\lim_{\epsilon \to 0} [p_i^{\epsilon}/q_i^{\epsilon}(u)] = 1 .$$

To prove this theorem we use the special representation for the stationary distribution of a finite Markov chain given in [FreW]. The required expression for the stationary probabilities can be written using the following $\{i\}$-graph construction.

Suppose we have a finite set $Z = (1, 2, \ldots, M)$. An $\{i\}$-graph is a directed acyclic graph with vertex set Z and a set of $M - 1$ edges, (k, ℓ), one incident out of each vertex $k \neq i$. The set of all $\{i\}$-graphs on Z will be denoted by $\mathbf{G}_Z(i)$; A_G will denote the set of edges in $G \in \mathbf{G}_Z(i)$.

Lemma 1 [FriW]. Consider a regular Markov chain with states $\{1, 2, \ldots, M\}$ and the transition matrix $P = \|p_{k\ell}\|_{k, \ell=1}^M$. Then the stationary distribution of this chain is a vector with components $p_i = \alpha_i / \sum\limits_{j=1}^M \alpha_j$, $i = 1, 2, \ldots, M$, where

$$(8) \qquad \alpha_i = \sum_{G \in \mathbf{G}_Z(i)} \prod_{(k, \ell) \in A_G} p_{k\ell} \, .$$

Proof of the theorem. For set U with values $U_1, U_2, \ldots, U_K$ define $d_U^\epsilon = \sum\limits_{i \in U} p_i^\epsilon$

and $e_i^\epsilon(U) = p_i^\epsilon / d_U^\epsilon$, $i \in U$. We introduce the stochastic matrix D^ϵ constructed from the matrix P^ϵ in the same way as Q^ϵ, except that instead of the vector $p^o(U)$ in (4) and (5), we use the vector $e^\epsilon(U) = (e_i^\epsilon(U); i \in U)$. The regularity of the chain with transition matrix D^ϵ follows directly from its construction and the regularity of the initial chain with transition matrix P^ϵ. It is easy to show that the vector $d^\epsilon = (d_{U_1}^\epsilon, \{p_i^\epsilon; i \in u_1\}, \ldots, d_{U_K}^\epsilon, \{p_i^\epsilon; i \in u_K\})$ satisfies the equation $d^\epsilon D^\epsilon = d^\epsilon$. Since its components are positive and sum to 1, d^ϵ is the vector of stationary probabilities for the matrix D^ϵ. Using (8) to calculate e_i^ϵ we have

$$(9) \qquad e_i^\epsilon(U) = \sum_{G \in \mathbf{G}_X(i)} \prod_{(k, \ell) \in A_G} p_{k\ell}^\epsilon \bigg/ \sum_{i \in U} \sum_{G \in \mathbf{G}_X(i)} \prod_{(k, \ell) \in A_G} p_{k\ell}^\epsilon \, .$$

With the representation (2) we collect in the numerator and denominator of (9) the coefficients of ϵ^m, where $m \geq 1$ is chosen as small as possible (in view of (2) this minimum value will be the same for all $i \in U$). We then reduce the fraction by the common factor. As a result we conclude

$$(10) \quad e_i^\epsilon(U) = \sum_{G \in \mathbf{G}_U(i)} \prod_{(k, \ell) \in A_G} p_{k\ell}^o \bigg/ \sum_{i \in U} \sum_{G \in \mathbf{G}_U(i)} \prod_{(k, \ell) \in A_G} p_{k\ell}^o + O(\epsilon) \, .$$
$$= p_i^o(U) + O(\epsilon) \, .$$

Further, calculating d^ϵ from (8) and taking (2) and (10) into account we obtain

$$(11) \qquad d^\epsilon = q^\epsilon + O(\epsilon) \mathbf{1}(n - L + K) \, .$$

Now (6) and (7) follow from (10) and (11) and the definitions of $e_i^\epsilon(U)$ and d_U^ϵ. ∎

The above theorem reduces the investigation of the limiting behavior of the stationary probability vector of the matrix P^ϵ to that of the stationary probability vector of Q^ϵ. Note that the order of Q^ϵ is $L-K$ less than the order of P^ϵ.

With the characterization of Lemma 1 and the following representation of the j-th component of q^ϵ

(12) $$q_j^\epsilon = \chi_j^\epsilon / (\chi_1^\epsilon + \ldots + \chi_{n-L+K}^\epsilon),$$

where the χ_i^ϵ, $1 \leqslant i \leqslant n-L+K$, are calculated from (8) with $P = Q^\epsilon$, it is easy to prove the following theorem.

Theorem 2. Suppose the matrix Q^ϵ is constructed from the nearly decomposable transition matrix P^ϵ, which satisfies condition (2). Then, for the K states $j = j_1, j_2, \ldots, j_K$ which correspond to the ergodic sets $U_1, U_2, \ldots, U_K$ of the chain with transition matrix P^o, we have

(13) $$\chi_j^\epsilon = \epsilon^{K-1} \sum_{i=0}^{k(j)} \alpha_{ji} \, \epsilon^i$$

where $\alpha_{j0} \neq 0$. For the remaining values $j \neq j_1, j_2, \ldots, j_K$ which correspond to the transient states of the chain with transition matrix P^o, we have

(14) $$\chi_j^\epsilon = \epsilon^K \sum_{i=0}^{k(j)} \beta_{ji} \epsilon^i,$$

where $\beta_{j0} \neq 0$.

With the states $j_1, j_2, \ldots, j_K$ introduced in Theorem 2 we can construct a regular Markov chain for which the vector $(q_j^*; j = j_1, j_2, \ldots, j_K)$, $q_j^* = \lim_{\epsilon \to 0} q_j^\epsilon$, will be the stationary probability vector. The subset of states $\{j_1, j_2, \ldots, j_K\}$ is denoted by C while the subset of remaining states of the chain with transition matrix Q^ϵ is denoted by H. For any two states $j_k, j_e \in C$ corresponding to the sets U and Y respectively, we consider a subgraph $G(U,Y)$ of the $\{j_e\}$-graph which consists of those edges $(j_k, i_1), (i_1, i_2), \ldots, (i_r, j_e)$ such that the intermediate states $i_1, \ldots, i_r$, if any, are in H, and the elements of Q^ϵ corresponding to transitions in the direction of each edge are positive.

We denote the product of the elements of Q^ϵ corresponding to the subgraph $G(U,Y)$ by $\pi(U,Y)$, while the set of all such subgraphs is denoted $\mathbf{G}(U,Y)$. In view of (4) and (5) and the fact that the matrices $R(u_j \times U_\ell)$ and $R(u_j \times u_\ell)$, $j = 1,2,\ldots,K$, $\ell = j, j+1,\ldots,K$, have only zero elements, $\pi(U,Y)$ can be rendered in the form

$$\pi(U,Y) = \epsilon\pi_1(U,Y),$$

where the product $\pi_1(U,Y)$ contains only cofactors expressed in terms of the matrices P^o and R and is not dependent on ϵ. We introduce a stochastic matrix S having nondiagonal elements

$$s(j_k, j_\ell) = \frac{1}{n} \sum_{G(U,Y)\,\in\,\mathbf{G}(U,Y)} \pi_1(U,Y).$$

The regularity of the chain with transition matrix S follows from its construction and from the regularity of the initial chain with transition matrix P^ϵ. The stationary probability vector of S is denoted by $\xi = (\xi(U_1),\ldots,\xi(U_K))$.

Theorem 3. Suppose the transition matrix P^ϵ of the nearly decomposable Markov chain satisfies condition (2). Then

(15)
$$\lim_{\epsilon \to 0} p_i^\epsilon = p_i^o(U)\xi(U)$$

for $i \in U$ and $U = U_1, U_2, \ldots, U_K$, and

(16)
$$\lim_{\epsilon \to 0} p_i^\epsilon = 0$$

for $i \in u$ and $u = u_1, u_2, \ldots, u_K$.

Proof. We apply Theorem 2 and examine $\xi(U)$, as obtained from (8), after the substitution $P = S$ and a reduction by a factor of $1/n$. Note that the matrix S is constructed in such a way that for $U = U_1, U_2, \ldots, U_K$ the numerator and denominator in the expression for $\xi(U)$ are identical to the coefficients of ϵ^{K-1}, where $K-1$ is the smallest power of ϵ in the numerator and denominator of the expression for $q^\epsilon(U)$ obtained from (12), i.e. $\lim_{\epsilon \to 0} q^\epsilon(U) = \xi(U)$. Hence (15) follows from (6). The second assertion in (16) follows from (7) after incorporating (12)-(14). ∎

We now make a few observations that refine the results.

Remark 1. Consider the matrix Q with nondiagonal elements

$$q(j_k, j_\ell) = ns(j_k, j_\ell)$$

and diagonal elements equal to minus the sum of the nondiagonal elements. It is easy to see that the vector ξ is uniquely defined by the equations

(17) $$\xi Q = 0(K), \quad \xi 1'(K) = 1$$

where $0(K)$ is a row vector of K zeros.

In the special case when the matrix P^ϵ is nearly completely decomposable, i.e. the limiting matrix P^o defines a Markov chain consisting only of ergodic sets, the matrix ϵQ is obtained by subtracting the unit matrix from the matrix defined by (4); the Simon-Ando approximation follows from (15) and (17) [CouL].

Remark 2. According to Lemma 1 the stationary distribution $(p_1^\epsilon, p_2^\epsilon, \ldots, p_n^\epsilon)$ of an arbitrary nearly decomposable Markov chain can be written as $p_i^\epsilon = \phi_i(\epsilon)/\psi_i(\epsilon)$, $i = 1, 2, \ldots, n$, where $\phi_i(\epsilon)$ and $\psi_i(\epsilon)$, by (8) and our assumptions regarding P^ϵ, are polynomials in ϵ. These polynomials can be represented in the form

$$\phi_i(\epsilon) = \epsilon^{\ell(i)} \sum_{j=0}^{n_1(i)} \alpha_{ij} \epsilon^j, \quad \psi_i(\epsilon) = \epsilon^{m(i)} \sum_{j=0}^{n_2(i)} \beta_{ij} \epsilon^j,$$

where the coefficients α_{i0} and β_{i0} are non-zero. Since $0 \leqslant p_i^\epsilon \leqslant 1$ for all $\epsilon > 0$, then $\ell(i) \geqslant m(i)$ and we can obtain the asymptotic expansions

(18) $$p_i^\epsilon = p_i^* + \epsilon \gamma_{i1} + \epsilon^2 \gamma_{i2} + \ldots, \quad i = 1, 2, \ldots, n.$$

Under the conditions of Theorem 3, it follows from (18), (15) and (16) (or from Theorem 1 and (12)-(14)) that

(19) $$p_i^\epsilon = \begin{cases} p_i^o(U)\xi(U) + O(\epsilon) & \text{for } i \in U, \ U = U_1, U_2, \ldots, U_K \\ O(\epsilon) & \text{for } i \in u, \ u = u_1, u_2, \ldots, u_K. \end{cases}$$

Note that the matrix Q^ϵ enables us to obtain not only p_i^ϵ, but all terms of the expansion (18). Recent results on error bounds for nearly completely decomposable Markov chains can be found in [CouS, Ste1, Ste2].

Remark 3. For the nearly decomposable Markov chain $\{x_t^\epsilon\}$ we may wish to consider the problem of finding the limiting probabilities

$$\lim_{\epsilon \to 0} Pr \left\{ x_t^\epsilon = j \mid x_0^\epsilon = i \right\} , \quad i, j = 1, 2, \ldots, n,$$

when $t = t(\epsilon) \to \infty$ as $\epsilon \to 0$. The following three cases have meaning:

$$a) \; \epsilon t(\epsilon) \to 0 \quad \text{as } \epsilon \to 0,$$
$$b) \; \epsilon t(\epsilon) \to \infty \quad \text{as } \epsilon \to 0,$$
$$c) \; \epsilon t(\epsilon) \to \tau \quad \text{as } \epsilon \to 0, \text{ where } 0 < \tau < \infty.$$

In case a)

$$\lim_{\epsilon \to 0} Pr \left\{ x_{t(\epsilon)}^\epsilon = j \mid x_0^\epsilon = i \right\} = \begin{cases} p_i^o(U) \text{ when } i, j \in U, \; U = U_1, \ldots, U_K \\ 0 \qquad \text{otherwise}. \end{cases}$$

In case b)

$$\lim_{\epsilon \to 0} Pr \left\{ x_{t(\epsilon)}^\epsilon = j \mid x_0^\epsilon = i \right\} = \lim_{\epsilon \to 0} p_j^\epsilon,$$

where the values of the latter limit are given by Theorem 3.

In case c) we confine ourselves to formulating the result for the nearly completely decomposable transition matrix P^ϵ. We consider the matrix Q introduced in Remark 1. For the nearly completely decomposable matrix P^ϵ the nondiagonal elements of the matrix Q corresponding to transitions between different ergodic sets U and Y are defined by

$$q_{U,Y} = p^o(U) R(U \times Y) 1'(n_Y),$$

whereas the diagonal elements are chosen so that the sum of the elements in each row is zero. For $Y = U_1, \ldots, U_K$, and any $U = U_1, \ldots, U_K$, let $\mu_{U,Y}(t)$ be an element of the row vector $\mu_U(t)$ of order K which is the solution of

$$\frac{d\mu_U(t)}{dt} = \mu_U(t) Q$$

with initial distribution $\mu_U(0) = (0, \ldots, 0,1,0, \ldots, 0)$, where the one is in the position corresponding to set U. Then when $i \in U$ and $j \in Y$

$$\lim_{\epsilon \to 0} Pr \left\{ x_{t(\epsilon)}^\epsilon = j \mid x_0^\epsilon = i \right\} = \mu_{U,Y}(\tau) p_j^o(Y) .$$

These probabilities are naturally called substationary.

The following two examples illustrate cases where we can simplify the calculation of the limiting stationary distribution of the nearly decomposable Markov chain.

Example 1. Consider the transition matrix of order n given by

$$P^\epsilon = \begin{bmatrix} P^o(U \times U) & 0 & 0 & 0 & 0 & 0 \\ P^o(u \times U) & P^o(u \times u) & 0 & 0 & 0 & 0 \\ 0 & 0 & P^o(V \times V) & 0 & 0 & 0 \\ 0 & 0 & P^o(v \times V) & P^o(v \times v) & 0 & 0 \\ 0 & 0 & 0 & 0 & P^o(W \times W) & 0 \\ 0 & 0 & 0 & 0 & P^o(w \times W) & P^o(w \times w) \end{bmatrix} + \epsilon R$$

where capital and small letters denote the respective ergodic and transient sets of the Markov chain with transition matrix P^o. Here, $p_{i,i-1}^o = 1$ for all $i \in u$, v and w. In the matrices $P^o(u \times U)$, $P^o(v \times V)$ and $P^o(w \times W)$ only the elements with the respective indices $(k_1, k_1 - 1)$, where $k_1 = n_U + 1$, $(k_2, k_2 - 1)$, where $k_2 = n_U + n_u + n_V + 1$, and $(k_3, k_3 - 1)$ where $k_3 = n - n_w$, are nonzero. The matrices $R(u \times X)$, $R(v \times X)$, $R(w \times X)$, $R(U \times (X \backslash w))$, $R(V \times (X \backslash u))$, and $R(W \times (X \backslash v))$ are zero and in the matrices $R(U \times w)$, $R(V \times u)$, and $R(W \times v)$ only the elements $r_{1n}, r_{\ell_1, \ell_1 - 1}$, $\ell_1 = n_U + n_u + 1$, and $r_{\ell_2, \ell_2 - 1}$, $\ell_2 = n_U + n_u + n_V + n_v + 1$, are nonzero. Such a matrix P^ϵ describes the behavior of a program with three localities U, V and W that alternate with linear (sequentially referenced) sets u, v and w. If the program begins to operate with pages belonging to locality U, then an exit from this locality can occur only after referencing page n in the linear set w. After sequential referencing of pages in w the program references page $n - n_w + 1$ belonging to locality W. Subsequent behavior of the program is described similarly with first, U replaced by W and w by v, and then W replaced by V and v by u. The first cycle is completed when the program returns to locality U.

The matrix Q^ϵ in this example has the following form:

$$
Q^\epsilon = \begin{bmatrix}
q^\epsilon_{11}\,0\,0 & & & & & & 0\ q^\epsilon_{1,n_u+n_v+n_w+3} \\
1\quad 0\,0 & & & & & & 0\,0 \\
0\quad 1\,0 & & \cdots & & & & 0\,0 \\
\vdots & & & & & & \\
0\cdots 0 & 1\,0 & 0 & & & & \\
0\cdots 0 & 0\,q^\epsilon_{k,k-1}\ q^\epsilon_{k,k} & & & & \vdots & \\
0\cdots 0 & 0\,0 & 1 & & & & \\
\vdots & & & & & & \\
0 & \cdots & & 0\,1\,0 & 0 & & \\
0 & \cdots & & 0\,0\,q^\epsilon_{\ell,\ell-1}\ q^\epsilon_{\ell,\ell} & & & \\
0 & \cdots & & 0\,0\,0 & 1 & & \\
\vdots & & & & & 1\,0\,0 & \\
0 & \cdots & & & & 0\,1\,0 &
\end{bmatrix}
$$

where

(20)
$$ q^\epsilon_{1,n_u+n_v+n_w+3} = \epsilon p^0_1(U)r_{1n} = \epsilon\beta_1 , $$

(21)
$$ q^\epsilon_{k,k-1} = \epsilon p^0_1(V)r_{\ell_1,\ell_1-1} = \epsilon\beta_2 , $$

(22)
$$ q^\epsilon_{\ell,\ell-1} = \epsilon p^0_1(W)r_{\ell_2,\ell_2-1} = \epsilon\beta_3 . $$

The matrix Q for Q^ϵ can be rendered as

$$
Q = \begin{bmatrix}
-q_{13} & 0 & q_{13} \\
q_{21} & -q_{21} & 0 \\
0 & q_{32} & -q_{32}
\end{bmatrix} ,
$$

where

$$q_{13} = \beta_1 \;, \; q_{21} = \beta_2 \;, \; q_{32} = \beta_3 \;.$$

To find the vector $\xi = (\xi(U),\xi(V),\xi(W))$ we solve (17) with the above matrix Q. We have

$$\xi(U) = \frac{\beta_2\beta_3}{\beta_1\beta_2 + \beta_1\beta_3 + \beta_2\beta_3}, \; \xi(V) = \frac{\beta_1\beta_3}{\beta_1\beta_2 + \beta_1\beta_3 + \beta_2\beta_3}, \; \xi(W) = \frac{\beta_1\beta_2}{\beta_1\beta_2 + \beta_1\beta_3 + \beta_2\beta_3} \,,$$

where β_1, β_2 and β_3 are found from (20)-(22).

These results can be generalized easily to the case of K sets $U_1, \ldots, U_K$ and K sets $u_1, \ldots, u_K$.

Example 2. Consider the transition matrix P^ϵ of order n, represented in the form

$$P^\epsilon = \begin{bmatrix} P^o(U \times U) \, 0 & & 0 \\ P^o(u \times U) & P^o(u \times u) \; P^o(u \times V) \\ 0 & 0 & P^o(V \times V) \end{bmatrix} + \epsilon R \;.$$

Here, the matrix $P^o(u \times X)$, with $X = U + u + V$ has a band structure, i.e. only the elements $p^o_{i,i-1}$, p^o_{ii} and $p^o_{i,i+1}$ are nonzero; in the matrix $R(U \times u)$ only one element $r_{k,k+1}$, $k = n_U$, is nonzero. All elements of the matrices $R(U \times V)$ and $R(V \times U)$ are zero, and in the matrix $R(V \times u)$ only the one element $r_{\ell+1,\ell}$, $\ell = n_U + n_u$, is nonzero. For this stochastic matrix, P^ϵ, the matrix Q^ϵ has the form

$$Q^\epsilon = \begin{bmatrix} q^\epsilon_{11} & q^\epsilon_{12} & 0 & 0 & & \cdots & & 0 \\ q^\epsilon_{21} & q^\epsilon_{22} & q^\epsilon_{23} & 0 & & \cdots & & 0 \\ & & \cdot & & \cdot & & & \\ & & & \cdot & & \cdot & & \\ & & & & \cdot & & & \\ 0 & \cdots & & & & q^\epsilon_{m-1,m-2} & q^\epsilon_{m-1,m-1} & q^\epsilon_{m-1,m} \\ 0 & \cdots & & & & 0 & q^\epsilon_{m,m-1} & q^\epsilon_{mm} \end{bmatrix}$$

where $q_{ij}^{\epsilon} = p_{k+i-1,k+j-1}^{o}$ with $2 \leqslant i \leqslant m-1$, $m = n_u + 2$, $q_{12}^{\epsilon} = \epsilon p_k^o(U) r_{k,k+1}$, and $q_{m,m-1}^{\epsilon} = \epsilon p_1^o(V) r_{\ell+1,\ell}$. The matrix Q for Q^{ϵ} has the form

$$
(23) \qquad\qquad Q = \begin{bmatrix} -q_{12} & q_{12} \\ q_{21} & -q_{21} \end{bmatrix},
$$

where

$$
q_{12} = p_k^o(U) r_{k,k+1} \, p_{k+1,k+2}^o \cdots p_{k+m-2,k+m-1}^o ,
$$

$$
q_{21} = p_1^o(V) r_{\ell,\ell+1} \, p_{\ell,\ell-1}^o \cdots p_{k+1,k}^o .
$$

Solving (17) for the matrix Q specified by (23), we obtain

$$
\xi(U) = \frac{q_{21}}{q_{12} + q_{21}} , \quad \xi(V) = \frac{q_{12}}{q_{12} + q_{21}} .
$$

These results can be generalized easily to an arbitrary number of sets $U_1, U_2, \ldots, U_K$ if the submatrix of P^{ϵ} corresponding to any three sets U_i, u_i and U_{i+1}, where $i = 1, 2, \ldots, K-1$, has the same structure as in the above example.

4. Working-Set Size Distributions

General Markov Model [HofT]. The working set (WS) and its size were introduced in Definition 5 of section 4.4. Explicit expressions were derived for the expected WS size and miss ratio in the IRM case. To generalize these results to a model of program behavior described by an ergodic Markov chain with transition matrix P, we define the reverse process of $\{x_t\}$ as that process corresponding to the evolution in reverse order of the original chain [KemS]. It is further assumed that the process $\{x_t\}$ is stationary, i.e. its initial distribution is its stationary distribution. This last assumption makes the reverse process a Markov chain with the transition matrix

$$
\hat{P} = [D(p)]^{-1} P \prime D(p),
$$

or termwise, $\hat{p}_{ij} = p_j p_{ji}/p_i$, where $p_i = Pr\{x_t = i\}$, $t \geqslant 0$, and $D(p)$ denotes the diagonal matrix with elements p_i, $i = 1, \ldots, n$. It is known [KemS] that the matrix $\hat{P}$ for the reverse chain possesses the same stationary distribution $(p_1, \ldots, p_n)$ as the direct chain.

The reverse chain is a natural tool for the study of working sets, since given x_t, it is $\hat{P}$ that governs WS dynamics directly, through a backward looking window of $\{x_t\}$.

Theorem 4. The steady state probability distribution of the working set size $S(T)$ is

$$(24)\ Pr\left\{S(T) = k\right\} = \sum_{\ell=0}^{k} (-1)^{k-\ell} \binom{n-\ell}{k-\ell} \sum_{1 \le j_1 < \ldots < j_\ell \le n}\ \sum_{i,j=1} \left(\hat{P}_{j_1} + \ldots + \hat{P}_{j_\ell}\right)^T_{ij}$$

where each of the n matrices $\hat{P}_j$ is the zero matrix except for column j, which is column j of $\hat{P}$.

Proof. The proof of this result is within easy reach of the following lemma.

Lemma 2. The conditional probability that the working-set size $S(t,T)$ at time t is k, given that page i is referenced at time $t+1$, is

$$Pr\left\{S(t,T) = k \,|\, x_{t+1} = i\right\} =$$

$$(25) \qquad = \sum_{\ell=1}^{k} (-1)^{k-\ell} \binom{n-\ell}{k-\ell} \sum_{1 \le j_1 < \ldots < j_\ell \le n}\ \sum_{j=1}^{n} \left(\hat{P}_{j_1} + \ldots + \hat{P}_{j_\ell}\right)^T_{ij}.$$

Proof. The argument is based on the observation that, given the state of the page reference process at time $t+1$, T transitions governed by $\hat{P}$ determine $W(t,T)$. Briefly, the argument runs as follows.

It is convenient to say that the event $\{S(t,T) = k\}$ occurs when the program avoids precisely $n-k$ pages in the interval $[t-T+1,t]$. Denote by $A_{I|i}$ the event: $x_{t-T+1}, \ldots, x_t$ avoids the set of pages I, given that $x_{t+1} = i$. Let $Q_{r|i}$ denote the sum $\sum_I P\{A_{I|i}\}$, the summation being over all sets I such that $|I| = r$. Then by the inclusion-exclusion principle [Fel]

(26) $$Pr\left\{S(t,T) = k \mid x_{t+1} = i\right\} = \sum_{m=0}^{k} (-1)^m \begin{bmatrix} n-k+m \\ m \end{bmatrix} Q_{n-k+m|i}$$

The probability that starting from i the next T references under $\hat{P}$ are all within a set of ℓ pages, J_ℓ, is given by

(27) $$\sum_{r=1}^{n} \left[\hat{P}_{j_1} + ... + \hat{P}_{j_\ell}\right]_{ir}^{T}, \quad J_\ell = \left\{j_1, \ldots, j_\ell\right\}.$$

From the definition of $Q_{r|i}$ and (27) we obtain for the value of $Q_{r|i}$

(28) $$Q_{r|i} = \sum_{J_{n-r}} \sum_{j=1}^{n} \left[\hat{P}_{j_1} + ... + \hat{P}_{j_{n-r}}\right]_{ij}^{T}, \quad J_{n-r} = \left\{j_1, \ldots, j_{n-r}\right\},$$

where the summation is over all sets of size $n - r$. Substitution of (28) into (26) and putting $k-m=\ell$ yields (25) (where we note that $Q_{n|i} = 0$), thus proving Lemma 2.

Finally, (24) is obtained by removing the conditioning on i. ∎

Using (24) the mean of the WS size can now be computed. As before denote the mean by $\omega(T)$, so that

$$\omega(T) = E[S(T)] = \sum_{k=1}^{n} kPr\{S(T) = k\}.$$

Theorem 5. The expected WS size is given by

(29) $$\omega(T) = n - \sum_{k=1}^{n} \sum_{i,j=1}^{n} p_i \left(\underline{\hat{P}}_k\right)_{ij}^{T},$$

where $\underline{\hat{P}}_k = \hat{P} - \hat{P}_k$, $k = 1, \ldots, n$.

Proof. First, we write

(30) $$\omega(T) = \sum_{k=1}^{n} kPr\{S(T) = k\} = \sum_{k=1}^{n} k \sum_{\ell=1}^{k} (-1)^{k-\ell} \begin{bmatrix} n-\ell \\ k-\ell \end{bmatrix} Q_\ell,$$

where

$$Q_\ell = \sum_{i,j=1}^{n} p_i \sum_{1 \leqslant j_1 < \ldots < j_\ell \leqslant n} \left[\hat{P}_{j_1} + \ldots + \hat{P}_{j_\ell} \right]_{ij}^{T} .$$

Now change the order of summation over k and ℓ in (30) and sum over $m = k - \ell$, $0 \leqslant m \leqslant n$. This yields

$$\omega(T) = \sum_{\ell=1}^{n} Q_\ell \sum_{m=0}^{n-\ell} (-1)^m \binom{n-\ell}{m} (m+\ell) .$$

Distributing the sum over m we have

$$\ell \sum_{m=0}^{n-\ell} (-1)^m \binom{n-\ell}{m} = \ell \delta_{\ell,n} , \quad \sum_{m=0}^{n-\ell} (-1)^m m \binom{n-\ell}{m} = -\delta_{\ell,n-1} ,$$

where $\delta_{i,j}$ is 1 when $i = j$ and is 0 otherwise. Therefore

$$\omega(T) = n Q_n - Q_{n-1} = n \sum_{i=1}^{n} p_i \sum_{j=1}^{n} \left[\hat{P} \right]_{ij}^{T}$$

$$- \sum_{i,j=1}^{n} p_i \sum_{1 \leqslant j_1 < \ldots < j_{n-1} \leqslant n} \left[\hat{P}_{j_1} + \ldots + \hat{P}_{j_{n-1}} \right]_{ij}^{T} = n - \sum_{i,j=1}^{n} p_i \sum_{k=1}^{n} \left[\hat{P}_k \right]_{ij}^{T} . \; \blacksquare$$

A related result that can be calculated is the expected miss ratio,

$$\zeta(T) = \lim_{t \to \infty} Pr\left\{ x_t \notin W(t-1,T) \right\} .$$

We need only observe that the present model satisfies the assumptions of [DenSc], viz. reference strings are infinitely long, the stochastic model is time-homogeneous, and the correlation between references x_t and x_{t+k} vanishes in the limit $k \to \infty$. By virtue of these properties the following relation holds:

(31)
$$\zeta(T) = \omega(T+1) - \omega(T) .$$

From (31) and (29) we have

$$(32) \qquad \varsigma(T) = \sum_{k=1}^{n} \sum_{i,j=1}^{n} p_i \left[\left(\hat{\underline{P}}_k \right)^T \left(I - \hat{\underline{P}}_k \right) \right]_{ij}$$

where I is the unit matrix.

The next result gives alternative expressions for $\omega(T)$ and $\varsigma(T)$. While the derivations are not so intuitive, the corresponding computations are more efficient.

Theorem 6. We have

$$(33) \qquad \omega(T) = \sum_{s=0}^{T-1} \sum_{j,k=1}^{n} p_k \left(\hat{\underline{P}}_k \right)_{kj}^{s},$$

$$(34) \qquad \varsigma(T) = \sum_{k,j=1}^{n} p_k \left(\hat{\underline{P}}_k \right)_{kj}^{T}.$$

Proof. Directly from (29) we have

$$(35) \qquad \omega(T) = n - \sum_{k,j=1}^{n} \theta_{jk}(T),$$

where $\theta_{jk}(T) = \sum_{i=1}^{n} p_i \left(\hat{\underline{P}}_k \right)_{ij}^{T}$. The $\theta_{jk}(T)$ satisfy the recurrence

$$\theta_{jk}(T) = \sum_{i=1}^{n} p_i \sum_{h=1}^{n} \left(\hat{\underline{P}}_k \right)_{ih} \left(\hat{\underline{P}}_k \right)_{hj}^{T-1}$$

$$= \sum_{h=1}^{n} \left\{ \sum_{i=1}^{n} p_i \hat{p}_{ih} - \sum_{i=1}^{n} p_i \left(\hat{P}_k \right)_{ih} \right\} \left(\hat{\underline{P}}_k \right)_{hj}^{T-1}$$

$$= \sum_{h=1}^{n} \left\{ p_h - \sum_{i=1}^{n} p_i p_{ik} \delta_{k,h} \right\} \left(\hat{\underline{P}}_k \right)_{hj}^{T-1}$$

$$= \theta_{jk}(T-1) - p_k \left(\hat{\underline{P}}_k \right)_{kj}^{T-1}.$$

This recurrence for $\theta_{jk}(T)$ has the initial value

$$\theta_{jk}(1) = p_j - p_k \delta_{j,k},$$

which yields the solution

$$\theta_{jk}(T) = p_j - p_k \sum_{s=0}^{T-1} \left[\hat{\underline{P}}_k\right]^s_{kj}.$$

Substituting into (35) we obtain

$$\omega(T) = n - \sum_{k,j=1}^{n} p_j + \sum_{k,j=1}^{n} p_k \sum_{s=0}^{T-1} \left[\hat{\underline{P}}_k\right]^s_{kj} - \sum_{s=0}^{T-1} \sum_{j,k=1}^{n} p_k \left[\hat{\underline{P}}_k\right]^s_{kj},$$

and therefore

$$\zeta(T) = \omega(T+1) - \omega(T) = \sum_{j,k=1}^{n} p_k \left[\hat{\underline{P}}_k\right]^T_{kj}. \quad \blacksquare$$

Remarks.

1. The evaluation of $\zeta(T)$ does not require all of $\left[\hat{\underline{P}}_k\right]^T$; for each k it suffices to evaluate only the kth row of these matrices via the relation, $row_k\left[\hat{\underline{P}}_k\right]^i = row_k\left[\hat{\underline{P}}_k\right]^{i-1}\hat{\underline{P}}_k$.

2. The last of the above relations suggests that we calculate expected WS sizes by first evaluating $\zeta(T)$ from (34) and then using $\omega(T+1) = \omega(T) + \zeta(T)$.

3. For the mean WS size we can write [DenSc]

$$\omega(T) = \sum_{s=0}^{T-1} \left[1 - \sum_{k=1}^{n} p_k F_k(s)\right],$$

where $F_k(\cdot)$ is the distribution function for the inter-reference interval for page k. Comparison with (33) yields

(36) $$F_k(s) = 1 - \sum_{j=1}^{n} \left[\hat{\underline{P}}_k\right]^s_{kj}.$$

Let us now evaluate the number of operations involved in computing the various expressions for $\omega(T)$, $\zeta(T)$, etc. In the usual way we limit ourselves to the number of additions and multiplications of elements of $\hat{P}$ and p. Equation (25) requires, for each k and i, $nT \sum_{\ell=1}^{k} \ell \binom{n}{\ell}$ operations, and for the complete (conditional) probability distribution we need $n^2 T(n+1)2^{n-2}$ operations for every i. Taking i into account introduces a factor of n. Therefore, the computation of the steady state distribution of working-set size by (24) requires about n times the number of operations needed for (25). A straightforward calculation of (29) requires $n^2(n-1)^2 T$ operations. (By a more elaborate calculation using $p=p\hat{P}$ explicitly, this figure can be reduced rather simply by a factor of n. The calculation is more complex to control, however.) By the procedure in Remark 1 of this section, (33) requires no more than $n^3 T$ operations, which is approximately the same as the more involved calculation of (29).

Nearly Decomposable Markov Model [Kog6]. We now develop an approximation method. Returning to the nearly decomposable model of program behavior described by the transition matrix P^ϵ of form (2), we introduce the notation

$$f^\epsilon(i,k,T) = Pr\left\{S^\epsilon(t,T) = k \,|\, x_{t+1}^\epsilon = i\right\},$$

$$f^o(U,i,k,T) = Pr\left\{S_U^o(t,T) = k \,|\, x_{t+1}^o = i\right\}$$

where the superscript ϵ or o indicates that a quantity corresponds to a Markov chain with the transition matrix P^ϵ or P^o, respectively, and where U is an arbitrary locality set of the program. Note that by virtue of (25) these conditional distributions do not depend on t. We also define the limits

$$f^\epsilon(k,T) = \lim_{t \to \infty} Pr\left\{S^\epsilon(t,T) = k\right\},$$

$$f^o(U,k,T) = \lim_{t \to \infty} Pr\left\{S_U^o(t,T) = k\right\}.$$

By the formula of total probability we have

$$Pr\left\{S^\epsilon(t,T) = k\right\} = \sum_{i \in X} Pr\left\{x^\epsilon_{t+1} = i\right\} f^\epsilon(i,k,T) =$$

$$(37) \qquad = \sum_U \sum_{i \in U} Pr\left\{x^\epsilon_{t+1} = i\right\} f^\epsilon(i,k,T) + \sum_{i \in H} Pr\left\{x^\epsilon_{t+1} = i\right\} f^\epsilon(i,k,T),$$

where H is the set of transient states of the chain with transition matrix P^o. Passing to the limit $t \to \infty$ in (37) we obtain

$$(38) \qquad f^\epsilon(k,T) = \sum_U \sum_{i \in U} p^\epsilon_i f^\epsilon(i,k,T) + \sum_{i \in H} p^\epsilon_i f^\epsilon(i,k,T).$$

We now obtain an approximate expression for $f^\epsilon(i,k,T)$ when $i \in U$ and $\epsilon \to 0$. If the transition matrix P^ϵ satisfies condition (2), then in view of its structure and (19), we have for $i \in U$

$$\hat{p}^\epsilon_{ij} = \frac{p^\epsilon_j p^\epsilon_{ji}}{p^\epsilon_i} = \begin{cases} p^o_j(U) p^o_{ji}/p^o_i(U) + O(\epsilon), & j \in U, \\ O(\epsilon), & j \notin U. \end{cases}$$

Hence, from the definition of the matrix $\hat{P}_j$ it follows that for $i \in U$ and $\epsilon \to 0$

$$(39) \qquad \hat{P}^\epsilon_j = \hat{P}^o_j(U \times U) + O(\epsilon) R_1,$$

where R_1 is an $n \times n$ matrix and its elements are bounded in absolute value by a constant independent of ϵ. Substituting (39) with $j = j_1, \ldots, j_\ell$ into (25) for $P = P^\epsilon$, we obtain for $i \in U$, fixed T and $\epsilon \to 0$

$$(40) \qquad f^\epsilon(i,k,T) = \begin{cases} f^o(U,i,k,T) + O(\epsilon), & k \leqslant n_U, \\ O(\epsilon), & k > n_U. \end{cases}$$

Now observe that

$$f^o(U,k,T) = \sum_{i \in U} p^o_i(U) f^o(U,i,k,T)$$

is the stationary distribution of the working-set size for the program behavior model of n_U pages, described by a Markov chain with transition matrix $P^o(U \times U)$. Thus, substituting (19) and (40) into (38) gives us the following result.

Theorem 7. Suppose program behavior is described by a nearly decomposable Markov chain with a transition matrix P^ϵ satisfying condition (2). Then we have for a stationary distribution of the working set size

$$(41) \qquad f^\epsilon(k,T) = \sum_U \xi(U)f^o(U,k,T) + O(\epsilon),$$

where the summation is over all localities U of the program and the vector $\xi = (\xi(U); U = U_1, \ldots, U_K)$ is determined by (17).

By (41) or by substituting (19) and (39) for p_i^ϵ and $\hat{P}_k^\epsilon$ into (29) and (32) with the assumptions of Theorem 7, it is easy to obtain approximations for the expected WS size and the WS miss ratio. Specifically, for ϵ small we have

$$(42) \qquad \omega^\epsilon(T) \approx \sum_U \xi(U)\omega_U^o(T),$$

$$(43) \qquad \zeta^\epsilon(T) \approx \sum_U \xi(U)\zeta_U^o(T),$$

where the summation is over all localities of the program, and the quantities $\omega_U^o(T)$ and $\zeta_U^o(T)$ are calculated from (29) and (32) or (33) and (34), respectively, for $n = n_U$ and $P = P^o(U \times U)$. The accuracy of the approximations (42) and (43) is $O(\epsilon)$.

We now make some observations concerning the use of the approximations.

Remark 4. If the program behavior inside each locality U is described by an independent or partial Markov reference model, then for $\omega_U^o(T)$ and $\zeta_U^o(T)$ we have simple explicit formulas (see (4.12), (4.13) and (4.10), (4.11)). In this case it remains only to obtain the vector ξ from (17). If the transition matrix P^ϵ has a structure of the type considered in the above examples, we can also write explicit formulas for the components of the vector ξ. Hence, the approximations (42) and

(43), unlike the exact expressions, have explicit forms in a greater number of cases.

Remark 5. An explicit expressioh for the stationary distribution of WS size is known only for the IRM [Van]

$$(44) \qquad f(k,T) = \sum_{\ell=1}^{k}(-1)^{k-\ell} \binom{n-\ell}{k-\ell} \sum_{1 \leqslant j_1 < ... < j_\ell \leqslant n} \left[p_{j_1}+...+p_{j_\ell} \right]^T$$

Equation (44) is readily obtained from (24) when we take into account the fact that for the IRM, $Pr\left\{ S(t,T) = k \mid x_{t+1} = i \right\} = Pr\left\{ S(t,T) = k \right\}$ and $\hat{p}_{ij} = p_j$.

This fact narrows the possibilities of obtaining explicit results from approximation (41). On the other hand, we recall that the number of operations required to calculate the stationary distribution of WS size increases exponentially as the number of program pages increases. Hence, in general, the savings in computation obtained from approximation (41) will increase very rapidly as the number of pages in the localities decreases.

Remark 6. Approximation (41) leads to approximate formulas similar to (42) for the higher moments of WS size. Finally, by means of (36) one can readily obtain an approximation for the distribution function of the inter-reference interval for any page $k \in U$.

5. Calculation of Miss Ratios

If the reference string $x_1, x_2, \ldots, x_t, \ldots$ is described by a regular Markov chain, then the pair (x_t, X_{t-1}^A), where $X_t^A = f(X_{t-1}^A, x_t)$ is the memory state at time t under replacement algorithm A, is a Markov chain for any Markovian algorithm A with finite memory (see Definition 3 in section 4.1). For simplicity we assume that the rule $f(\cdot, \cdot)$ is deterministic. If the chain $\{(x_t, X_{t-1}^A)\}$ is ergodic then the miss ratio $F(A)$ can be defined as

$$(45) \qquad F(A) = F_m(A) = \lim_{t \to \infty} Pr\{x_t \notin X_{t-1}^A\} .$$

If the sequence of memory states $\{X_t^A\}$ also forms a Markov chain as in the IRM, then (45) can be calculated from (4.34). In general, to calculate (45) we must find the stationary probability vector $\{q^A(i,s)\}$, where $i \in X$ and $s \in \{s\}^A$ is a memory state of the chain $\{(x_t, X_{t-1}^A)\}$ with transition probabilities

(46) $p^A((i,s),(j,r)) = \begin{cases} p_{ij} & \text{if } i \in s \text{ and } r = s \text{ or } i \notin s \text{ and } r = f(s,i), \\ 0 & \text{otherwise.} \end{cases}$

Thus, $\{q^A(i, s)\}$ is the invariant probability vector of the transition matrix with elements (46), and by the ergodic theorem and the formula of total probability we have

(47) $$F(A) = \sum_{(i,s)} q^A(i,s)[1 - \sum_{j \notin f(s,i)} p_{ij}] .$$

The computational complexity of (47) is overwhelmed by the enormous size, $m!\binom{n}{m}$, of the state space $\{(i,s)\}^A$. The following theorem provides an approximation for the miss ratio under the assumption of near decomposability, which can considerably reduce the complexity of the calculation.

Theorem 8. Suppose program behavior is described by a nearly decomposable Markov chain whose transition matrix P^ϵ satisfies condition (2). Then, for the Markovian replacement algorithm A with finite memory, as $\epsilon \to 0$

(48) $$F_m^\epsilon(A) = \sum_U \xi(U) F_m^0(U,A) + O(\epsilon) ,$$

where $F_m^0(U,A)$ is the miss ratio for the Markov model with page set U and transition matrix $P^0(U \times U)$, where the vector ξ with components $\xi(U)$, $U = U_1, \ldots, U_K$, is found from (17), and where the summation is carried out over all localities U of the program.

Proof. It follows from (46) that to the nearly decomposable Markov chain $\{x_t^\epsilon\}$ there corresponds the nearly decomposable Markov chain $\{x_t^\epsilon, X_{t-1}^{\epsilon,A}\}$. (Since replacement of the pages of a locality in main memory does not occur immediately after a page of a new locality is referenced, the Markov chain $\{x_t^\epsilon, X_{t-1}^{\epsilon,A}\}$ will be nearly decomposable even when the chain $\{x_t^\epsilon\}$ is nearly completely decomposable.) We will denote its transition matrix by Ω^ϵ. It is easy to see that there is a one-to-one correspondence between the ergodic sets of the chains with transition matrices P^o and Ω^o. In the latter, let Δ be the set of all possible pairs $(i, s(U))$, where $i \in U$ and $s(U)$ is the set of m distinct pages in U. Then Δ corresponds to the set U in the chain with transition matrix P^o. (For simplicity we will assume that $n_U \geqslant m$ for all U). In view of the near decomposability of the matrix Ω^ϵ, we obtain from (47) that as $\epsilon \to 0$

$$F_m^\epsilon(A) = \sum_\Delta \xi_\Omega(\Delta) F_m^o(\Delta, A) + O(\epsilon),$$

where the summation is carried out over all ergodic sets, Δ, of the chain with transition matrix Ω^o, where $F_m^o(\Delta, A) = F_m^o(U, A)$, and where

$$\xi_\Omega(\Delta) = \lim_{\epsilon \to 0} \lim_{t \to \infty} Pr\left[(x_t^\epsilon, X_{t-1}^{\epsilon, A}) \in \Delta\right].$$

Hence, to prove the theorem it remains to show that

(49) $$\xi_\Omega(\Delta) = \xi(U) .$$

First, we write

(50) $$Pr\left\{x_t^\epsilon = i\right\} = \sum_s Pr\left\{(x_t^\epsilon, X_{t-1}^{\epsilon, A}) = (i, s)\right\},$$

where the summation is over all states $s \in \{s\}^A$. If we suppose that $i \in U$, then it is easily seen that

$$\lim_{\epsilon \to 0} \lim_{t \to \infty} Pr\left\{(x_t^\epsilon, X_{t-1}^{\epsilon, A}) = (i, s)\right\} = 0$$

when $s \neq s(U)$. Hence, from (46) it follows that when $P = P^\epsilon$,

(51) $$\lim_{\epsilon \to 0} \lim_{t \to \infty} Pr\left\{x_t^\epsilon = i\right\} = \lim_{\epsilon \to 0} \lim_{t \to \infty} \sum_{s(U)} Pr\left\{(x_t^\epsilon, X_{t-1}^{\epsilon, A}) = (i, s(U))\right\}$$

Note that

(52) $$\xi(U) = \lim_{\epsilon \to 0} \lim_{t \to \infty} \sum_{i \in U} Pr\left\{x_t^\epsilon = i\right\}$$

and

(53) $$\xi_\Omega(\Delta) = \lim_{\epsilon \to 0} \lim_{t \to \infty} \sum_i \sum_{Us(U)} Pr\left\{(x_t^\epsilon, X_{t-1}^{\epsilon, A}) = (i, s(U))\right\}.$$

Thus, (49) follows from (50)-(53), and the proof of the theorem is complete. ■

In the next two subsections we present methods for efficient calculation of the miss ratio within a locality for the replacement algorithms, LRU and FIFO. These methods, together with approximate formula (48), extend the number of computable cases.

Efficient Calculation of the LRU Miss Ratio [Tze]. Let the reference string $x_1, x_2, \ldots, x_t, \ldots$ be described by a regular Markov chain with the transition matrix $P = ||p_{ij}||_{i,j=1}^n$. According to (45) the LRU miss ratio is

$$F_m(\text{LRU}) = \lim_{t \to \infty} Pr\left\{x_t \notin X_{t-1}^{LRU}\right\}$$

(54)

$$= \sum_{i=1}^n \left[\lim_{t \to \infty} Pr\left\{x_t \notin X_{t-1}^{LRU} \mid x_t = i\right\}\right] p_i,$$

where the latter equality is a consequence of the formula of total probability.

Next, observe that the parenthesized expression in the right-hand side of (54) is in fact the probability of first return to state i along a path in which at least m distinct pages are referenced. This follows directly from the structure of the algorithm LRU, as well as from renewal properties of the Markov chain $\{x_t\}$.

We can make this observation more precise, as follows. Consider a sample path $x_1, x_2, \ldots, x_T$ such that $x_1 = x_T = i$ and $x_t \neq i$ for all $1 < t < T$. The length of this path is defined to be $|L|$, where $L = \{x_2, x_3, \ldots, x_{T-1}\}$. It is proved in [Tze] that the probability, $\pi_{ii}(k)$ of the first return to state i along a path of length k is given by

$$(55) \qquad \pi_{ii}(k) = \sum_{r=1}^{k}(-1)^{k-r}\binom{n-r-1}{k-r}\sum_{J_r \subset X\backslash i}\left[P_{J_r}^k(I-P_{J_r})^{-1}P_i\right]_{ii},$$

where I is the unit matrix, $J_r = \{j_1, \ldots, j_r\}$ and P_{J_r} designates the algebraic sum of the r matrices $P_{j_1}, \ldots, P_{j_r}$, introduced in section 4. Thus

$$F_m(\text{LRU}) = \sum_{i=1}^{n}p_i\sum_{k=m}^{n-1}\pi_{ii}(k)$$

(56)

$$= \sum_{i=1}^{n}p_i\sum_{k=m}^{n-1}\sum_{r=1}^{k}(-1)^{k-r}\binom{n-r-1}{k-r}\sum_{J_r \subset X\backslash i}\left[P_{J_r}^k(I-P_{J_r})^{-1}P_i\right]_{ii}.$$

From (56) it follows that $F_m(\text{LRU}) \geqslant F_{m+1}(\text{LRU})$. An iterative computational procedure for obtaining $F_m(\text{LRU})$ for all $m = 1,2,\ldots,n$ is suggested by the recurrence

$$F_n(\text{LRU}) = 0,$$

$$(57) \qquad F_m(\text{LRU}) = F_{m+1}(\text{LRU}) + \sum_{i=1}^{n}p_i\pi_{ii}(m), \quad m = n-1,n-2,\ldots,1.$$

A procedure for computing the $\pi_{ii}(k)$, as defined in (55), is expressed below in ALGOL-like notation in order to simplify the problem of counting the necessary operations (see the bracketed counts given on the right).

for $i=1$ *until* n *do*	[n times]
begin	
for $m=1$ *until* $n-1$ *do*	[$n-1$ times]
begin	
for all tuples J_m *do*	[$\binom{n-1}{m}$ times]
begin	
compute the m significant	
columns	
$j_1,\ldots,j_m$ of $(I-P_{J_m})^{-1}$;	[$m^3/3$ multiplications]
postmultiply by P_i	[m^2 multiplications]
compute $P_{J_m}^m$	[$(m-1)m^2$ multiplications]

    for r=m until n−1 do   [n−m times]
       begin
       compute $P_{J_m}^r$ (given $P_{J_m}^{r-1}$);   [m^2 multiplications]
       compute the i-th row of
       $[P_{J_m}^r (I - P_{J_m})^{-1} P_i]$;   [$m$ multiplications]
       end;
     end;
   end;
 end

Inspection of the algorithm shows that the number of multiplications required is

$$n \sum_{m=1}^{n-1} \binom{n-1}{m} \left[(m^3/3) + m^2 + (m-1)m^2 + (n-m)(m^2 + m) \right]$$

(58)

$$= 2^{n-4} \left[\frac{7n^4}{3} + O(n^3) \right] .$$

An additional $O(n^3)$ multiplications are required when performing the iterative procedure of (57). This produces no change in (58), which therefore describes the complexity of the entire calculation. The incorporation of (57) into the calculation of miss ratios creates substantial improvements over earlier approaches; with the same investment in computer time this method can accommodate problem sizes (values of n) almost double those handled by Glowacki's formula [Glo], which is based on (47), and almost four times those treated by the method of Franklin and Gupta [FraG].

In [Tze] the path-length approach is also applied to the calculation of the miss ratio for the partially preloaded LRU algorithm (see Definition 7 in section 4.6).

Efficient Calculation of FIFO and LRU Miss Ratios in the IRM [FagP]. We begin by presenting an efficient, stable algorithm for evaluating the FIFO miss ratio (see (4.37)). We have

(59)
$$F_m(\text{FIFO}) = \frac{\displaystyle\sum_{(i_1,\ldots,i_m)} p_{i_1} \cdots p_{i_m} \sum_{j=m+1}^{n} p_{i_j}}{\displaystyle\sum_{(i_1,\ldots,i_m)} p_{i_1} \cdots p_{i_m}}$$

in the IRM as derived in section 4.6. The sums in (59) are taken over all m-tuples $(i_1, \ldots, i_m)$ such that $i_j \neq i_\ell$ if $j \neq \ell$.

Let $p_1, \ldots, p_n$ be a fixed but arbitrary ordering of the n page probabilities. For positive integers r and k, $k \leqslant n$, define

$$(60) \qquad \sum(r,k) = \sum_{\Lambda_{rk}} p_{i_1} p_{i_2} \cdots p_{i_r} \, ,$$

where Λ_{rk} is comprised of all r-element subsets $\{i_1, \ldots, i_r\}$ of $\{1, \ldots, k\}$. In other words, for each r-element subset of the first k probabilities $p_1, \ldots, p_k$, there is in $\sum(r,k)$ a term that is the product of these r probabilities. We adopt the usual convention that an empty sum is 0; hence, $\sum(r,k) = 0$ if $r > k$.

From (4.41) and (60) we have

$$(61) \qquad F_m(\text{FIFO}) = \frac{(m+1) \sum(m+1,n)}{\sum(m,n)} .$$

To calculate (61) both efficiently and stably, we make use of the following recurrence relation for $\sum(r,k)$ when $r > 1$ and $k > 1$:

$$(62) \qquad \sum(r,k) = \sum(r, k-1) + p_k \sum(r-1, k-1) .$$

In (62), $\sum(r, k-1)$ is the sum of those terms in $\sum(r, k)$ which do not have p_k as a factor, and $p_k \sum(r-1, k-1)$ is the sum of the remaining terms.

In order to determine (61) we need to calculate the ratio $\sum(m+1,n)/\sum(m,n)$, but not $\sum(m+1,n)$ and $\sum(m,n)$ individually. For this purpose let

$$F(r,k) = \sum(r,k)/\sum(r-1,k), \quad 1 \leqslant r \leqslant m+1, \ 1 \leqslant m \leqslant n .$$

Empty products are defined to have the value 1, and we define $\sum(0,k)$ to be 1 for all k. Hence, $F(1,k) = \sum(1,k) = p_1 + \ldots + p_k$, and from (61) the FIFO miss ratio becomes $(m+1) F(m+1,n)$. We can derive a recurrence for F directly and avoid the numerical difficulties created by (62). To this end, note first that if $r > 1$ and $k > 1$, then

(63)
$$F(r,k) = \frac{\sum(r,k)}{\sum(r-1,k)}$$

$$= \frac{\sum(r,k)/\sum(r-1,k-1)}{\sum(r-1,k)/\sum(r-2,k-1)} \cdot \frac{\sum(r-1,k-1)}{\sum(r-2,k-1)} \cdot$$

Dividing both sides of (62) by $\sum(r-1,k-1)$ produces

(64)
$$\sum(r,k)/\sum(r-1,k-1) = \sum(r,k-1)/\sum(r-1,k-1) + p_k$$

$$= F(r,k-1) + p_k .$$

By (64) we can replace $\sum(r,k)/\sum(r-1,k-1)$ in (63) by $F(r,k-1) + p_k$, and similarly, we can replace $\sum(r-1,k)/\sum(r-2,k-1)$ in (63) by $F(r-1,k-1) + p_k$; we obtain thereby the recurrence

(65) $$F(r,k) = \frac{F(r,k-1) + p_k}{F(r-1,k-1) + p_k} F(r-1,k-1), \ 1 \leqslant r \leqslant m+1, \ 1 \leqslant k \leqslant n .$$

From this recurrence we can compute the matrix of values, $F(r,k)$, starting from the boundary conditions

(66)
$$F(r,1) = \begin{cases} p_1, \ r=1 \\ 0, \ 2 \leqslant r \leqslant m+1, \end{cases}$$

$$F(1,k) = p_1 + ... + p_k, \ 1 \leqslant k \leqslant n .$$

One way to calculate F is to initialize the first column and row with the values for $F(r,1)$ and $F(1,k)$ in (66), and then to calculate the entries for $2 \leqslant r \leqslant m+1$ and $2 \leqslant k \leqslant n$, column by column, by means of (65). Then the FIFO miss ratio with capacity m is $(m+1)F(m+1,n)$.

This algorithm requires approximately 4nm additions, multiplications and divisions. Observe that it calculates F_1 (FIFO), . . . , F_{m-1}(FIFO) in the process of determining F_m (FIFO). A proof of its numerical stability is sketched in [FagP].

We turn now to an efficient method for obtaining an unbiased estimate of the expected LRU miss ratio given by

$$(67) \qquad F_m(\text{LRU}) = \sum_{(i_1, \ldots, i_m)} \frac{p_{i_1} \cdots p_{i_m}(1-p_{i_1}-\ldots-p_{i_m})}{(1-p_{i_1}) \cdots (1-p_{i_1}-\ldots-p_{i_{m-1}})}$$

in the IRM.

Consider the following experiment, which involves drawing integers (pages) from a set $X = \{1,2,\ldots,n\}$ without replacement. We say that integer i has weight p_i, $i = 1,\ldots,n$, where $\{p_1,\ldots,p_n\}$ is the page reference probability distribution. Select one integer from X in such a way that a given integer is selected with a probability equal to its weight. Thus, i is selected with probability p_i. Assuming that i_1 was selected, renormalize the weights of the remaining $(n-1)$ integers so that their sum is 1. Thus, the weight of j is now $p_j/(1-p_{i_1})$ for $j \neq i_1$. Now select an integer from $X\backslash\{i_1\}$, where once again a given integer is selected with a probability equal to its current weight. Assuming that i_2 was selected, renormalize the weights of the remaining $n-2$ integers so that their sum is 1, and continue this selection/renormalization process until m integers have been selected. Observe that with probability

$$p_{i_1} \frac{p_{i_2}}{1-p_i} \frac{p_{i_3}}{1-p_{i_1}-p_{i_2}} \cdots \frac{p_{i_m}}{1-p_{i_1}-\ldots-p_{i_{m-1}}},$$

i_1 was selected first, i_2 was selected second ,..., and i_m was selected last; in this event $\hat{F}_m(\text{LRU}) = 1-p_{i_1}-\ldots-p_{i_m}$ estimates (67). Therefore, the expected value over all selection sequences of the random variable $\hat{F}_m(\text{LRU})$ is given by (67); that is, $\hat{F}_m(\text{LRU})$ is an unbiased estimate of (67).

Examples

1. The experiment just described is easily programmed on a computer. As an example of the estimates obtainable in this way, we now describe a numerical example using Zipf's law [Zip, Knu p. 397]. Here, the probability p_i of referencing the i-th most frequently referenced page is

$$p_i = \frac{k}{i^\theta}, \quad 1 \leqslant i \leqslant n,$$

where $\theta > 0$ is the skewness parameter and k is chosen so that $\sum\limits_{i=1}^{n} p_i = 1$. In particular, for $\theta \approx 0.86$ this distribution gives the well-known "80/20 law" [Knu] which states that 80% of the references to a file occur to only 20% of the file, and 80% of these references occur to only 20% of the previous 20% of the file, and so on.

In our specific example, $\theta = 0.5$ was chosen for the skewness, $n = 100$ for the number of pages, and $m = 30$ for the capacity. The experiment was replicated 100 times to produce the estimates $\hat{F}^{(i)}$, $1 \leqslant i \leqslant 100$, whose average, $\bar{F}$, was then computed. Note that $\bar{F}$ is an unbiased estimate of (67) since each of the $\hat{F}^{(i)}$ is. For the given parameters, $\bar{F} = 0.6119$ rounded to 4 decimal places was found.

To determine how much confidence should be placed in this estimate, we calculated several other related statistical quantities. For L replications of the experiment let

$$\delta = \left[\frac{1}{L-1} \sum_{i=1}^{L} (\hat{F}^{(i)} - \bar{F})^2 \right]^{\frac{1}{2}} .$$

Then δ and $y = \delta/\sqrt{L}$ are unbiased estimates of the standard deviation and the standard deviation of the mean, respectively. In our example, where $L = 100$, δ turned out to be 0.0301, and so y was 0.0030. By the central limit theorem the normal approximation is very good for our sample of size 100. (The normal approximation was justified in the present example by the Kolmogorov-Smirnov test.) Under this approximation we know that an approximate 95% confidence interval for the sample mean is given by $\bar{F} \pm 2y$. Thus, with approximately 95% confidence we can say that the LRU miss ratio with parameters $n = 100$, $m = 30$ and Zipf's law with $\theta = 0.5$ is 0.6119 ± 0.0060.

2. As a further demonstration of the power of current techniques, including those developed in this section for the IRM, Table 1 presents a family of examples. In each case we assume Zipf's law with skewness $\theta = 0.5$. We vary the number n of pages, and we also vary the memory capacity m in such a way that the normalized capacity, m/n, is 0.3. All values are rounded to four decimal places. We include not only the expected FIFO and LRU miss ratios, but also the expected WS and

A_0 miss ratios. In the WS case, the window size T is chosen in such a way that m is the expected working-set size.

The fact that the LRU, WS and A_0 miss ratios have limiting values (as in Table 1) is proved in the next section, where closed-form formulas are exhibited for these limits. Further, it will be shown there that the limits in the LRU and WS cases are the same. (In the case of Table 1, the common limit is 0.5701). It is an open problem as to whether there is a limiting value for the FIFO miss ratio.

A minor technical comment on the LRU calculations in Table 1 should be made. Except for the $n = 10$ case, where formula (4.35) was used, the interval given in the LRU column is approximately a 95% confidence interval. For the $n = 100$ and $n = 1000$ cases, the experiment described in the previous subsection was performed $L = 100$ times. For $n = 10000$ only 30 iterations of the experiment were made, because of the great amount of paging that occurs when dealing with very large vectors.

Table 1

Miss ratios. Zipf's Law, skewness $\theta = 0.5$, normalized capacity $m/n = 0.3$

		FIFO	LRU	WS	A_0
$n = 10$,	$m = 3$	0.6660	0.6607	0.6599	0.5741
$n = 100$,	$m = 30$	0.6304	0.6119 ± 0.0060	0.6096	0.4870
$n = 1,000$,	$m = 300$	0.6091	0.5827 ± 0.0017	0.5831	0.4629
$n = 10,000$,	$m = 3,000$	0.6007	0.5748 ± 0.0010	0.5742	0.4556
Limiting value		?	0.5701	0.5701	0.4523

Asymptotic LRU Miss Ratios. We would expect the formulas to simplify as n increases and as the assigned memory size $m = m(n)$ and the probability distributions are modified accordingly. In what follows we see how this occurs for results on the asymptotic behavior of the LRU miss ratio. First, we give the following canonical method of determining the probabilities $\{p_1^{(n)}, p_2^{(n)}, ..., p_n^{(n)}\}$ for each n.

Let G be a continuously differentiable, monotonically increasing function over the interval $[0, 1]$ such that $G(0) = 0$ and $G(1) = 1$. For each positive integer n the probability distribution $\{p_1^{(n)},..., p_n^{(n)}\}$ is determined by

$$p_i^{(n)} = G(i/n) - G((i-1)/n) , \quad 1 \leqslant i \leqslant n .$$

Clearly $p_i^{(n)} > 0$ for each i and $\sum_{i=1}^{n} p_i^{(n)} = 1$. When n is understood in context, the superscript in $p_i^{(n)}$ will be omitted.

Let β_0 be a real number satisfying $0 \leqslant \beta_0 \leqslant 1$. The number β_0 is taken as approximately equal to m/n, where m is the memory size assigned to the program when the LRU algorithm is used. We denote by $F_{\lfloor \beta_0 n \rfloor}$ (LRU) the LRU miss ratio for the probability distribution $\{p_1,..., p_n\}$ determined by $G(x)$ and the memory size $m = \lfloor \beta_0 n \rfloor$. It can be shown [Fag] that

$$\lim_{n \to \infty} F_{\lfloor \beta_0 n \rfloor}^{(n)} (\text{LRU}) = K(\beta_0) \equiv \int_0^1 G'(x) e^{-\tau_0 G'(x)} dx ,$$

where the parameter $\tau_0 = \tau_0(\beta_0)$, $0 \leqslant \tau_0 \leqslant \infty$, is defined by the equation

(68) $$\beta_0 = 1 - \int_0^1 e^{-\tau_0 G'(x)} dx .$$

It can be proved that $K(\beta_0)$ is also the limiting value (as $n \to \infty$) of the working-set miss ratio when the expected working set size $\omega(T) = \lfloor \beta_0 n \rfloor$. In this case the following very simple formula gives the asymptotic value of the LRU miss ratio as the number of pages becomes large (see (4.13))

$$F_{\lfloor \beta_0 n \rfloor}^{(n)} (\text{LRU}) \approx \varsigma(T) = \sum_{i=1}^{n} p_i (1-p_i)^T$$

where the working-set parameter T is defined by

$$\omega(T) = \sum_{i=1}^{n} (1-(1-p_i)^T) = \lfloor \beta_0 n \rfloor .$$

These results can be extended to the semi-Markov model by means of (4.48). We denote by $\varsigma_{sM}(T)$ and $\varsigma_{IR}(T)$ the expected working-set miss ratios with parameter T in the semi-Markov model and the corresponding IRM, respectively.

Let the expected working-set size with parameter T for the IRM be

$$\omega_{IR}(T) = n - \sum_{i=1}^{n} p_i(1-p_i)^T = \lfloor \beta_0 n \rfloor \, ,$$

where n is fairly large. We then get the following approximate formula for the semi-Markov model with the same value of T:

$$F_{\lfloor \beta_0 n \rfloor}(LRU) \approx \zeta_{sM}(T) \approx \zeta_{IR}(T) / \sum_{i=1}^{n} p_i v_i$$

$$= \sum_{i=1}^{n} p_i(1-p_i)^T / \sum_{i=1}^{n} p_i v_i \, ,$$

where v_i is the mean number of repeated references to page i, assuming that after the last of these the reference to the next page occurs in accordance with the probability distribution $\{p_1, p_2,..., p_n\}$.

These results are illustrated by two examples [Fag].

Example 1. Consider again Zipf's law, where the probability p_i that page i is referenced is

$$p_i = \frac{k}{i^\theta},$$

with θ a positive constant and k chosen so that $\sum_{i=1}^{n} p_i = 1$. It can be shown that with $0 \leqslant \theta \leqslant 1$ it is sufficient to put $G(x) = x^{1-\theta}$, since for $n \to \infty$

$$\sum_{i=1}^{\lfloor \beta_0 n \rfloor} p_i^{(n)} \to \beta_0^{1-\theta}$$

for any $0 \leqslant \theta \leqslant 1$ and $0 < \beta_0 < 1$.

If $G(x) = x^{1-\theta}$, then (68) takes the following form in the definition of τ_o:

$$\beta_0 = 1 - \int_0^1 e^{-\tau_0(1-\theta)x^{-\theta}} dx$$

and

$$\lim_{n \to \infty} F_{\lfloor \beta_0 n \rfloor}^{(n)}(LRU) = K(\beta_0) = \int_0^1 (1-\theta)x^{-\theta} e^{-\tau_0(1-\theta)x^{-\theta}} dx \, .$$

Table 2 gives numerical results for $\theta = 0.5$ and $\beta_0 = 0.6$. In this case $\tau_0 = 1.1403$ and $K(\beta_0) = 0.2902$. The expected working-set miss ratio $\zeta(T)$ is calculated with that value of T for which the expected working-set size is $\omega(T) = 0.6\, n$. Note that (4.35) has over 30,000 terms and requires over 300,000 multiplications/divisions to calculate $F_m(\text{LRU})$ for $n = 10$ and $m = 6$; when $n = 100$ and $m = 60$, direct computation by (4.35) is clearly out of the question.

Example 2. Consider the arithmetic distribution $p_i = k(a+ib)$, where a and b are positive constants and the constant $k = k(n)$ is chosen from the normalization condition $\sum\limits_{i=1}^{n} p_i^{(n)} = 1$. It is straightforward to check that the quantity $p_1^{(n)} + \ldots + p_{\lfloor \beta_0 n \rfloor}^{(n)}$ converges to β_0^2 as $n \to \infty$ independently of a and b, which means that the arithmetic distribution corresponds to the function $G(x) = x^2$. Therefore, τ_0 is determined from

$$\beta_0 = 1 - \int\limits_0^1 e^{-2\tau_0 x} dx = \frac{2\tau_0 + e^{-2\tau_0} - 1}{2\tau_0},$$

with the limit

$$\lim_{n \to \infty} F_{\lfloor \beta_0 n \rfloor}^{(n)}(\text{LRU}) = K(\beta_0) = \int\limits_0^1 2x e^{-2\tau_0 x} dx = \frac{1 - e^{2\tau_0}(1-2\tau_0)}{2\tau_0^2}.$$

We end this section with some comments on the limiting behavior of the OPT miss ratio $F_{\lfloor \beta_0 n \rfloor}(\text{OPT})$ for $\beta_0 \to 1$ [Fag]. According to this algorithm, main memory always contains the $m-1$ pages with the highest reference probabilities. Therefore, (4.46) implies that when $G(x)$ is concave (i.e. when the derivative $G'(x)$ decreases monotonically), $F_{\lfloor \beta_0 n \rfloor}^{(n)}(\text{OPT}) \to 1 - G(\beta_0)$ as $n \to \infty$; if $G(x)$ is convex, then $F_{\lfloor \beta_0 n \rfloor}^{(n)}(\text{OPT}) \to G(1-\beta_0)$ as $n \to \infty$.

One might expect the ratio $F_{\lfloor \beta_0 n \rfloor}(\text{LRU}) / F_{\lfloor \beta_0 n \rfloor}(\text{OPT})$ to be close to one for n large and $\beta_0 \to 1$. However, this is not necessarily the case. With the arithmetic distribution, it is not difficult to show that as $\beta_0 \to 1$ (i.e. as $\tau_0 \to \infty$), $F_{\lfloor \beta_0 n \rfloor}^{(n)}(\text{LRU}) \approx 2(1-\beta_0)^2$, whereas $F_{\lfloor \beta_0 n \rfloor}(\text{OPT}) \approx 1 - \beta_0^2$. Thus, the ratio $F_{\lfloor \beta_0 n \rfloor}(\text{LRU}) / F_{\lfloor \beta_0 n \rfloor}(\text{OPT})$ is close to 2 for n large and β_0 close to 1.

In [Fag] an asymptotic formula is also derived for the miss ratio under VMIN, the optimal variable space replacement algorithm.

Table 2

Miss ratios, Zipf's Law, skewness $\theta = 0.5$, parameter $\beta_0 = 0.6$

n	$\omega(T) = \lfloor \beta_0 n \rfloor$	$\zeta^{(n)}(T)$	$F_{\lfloor \beta_0 n \rfloor}^{(n)}$ (LRU)
10	6	0.3492	0.3518
100	60	0.3109	
1000	600	0.2968	
10 000	6000	0.2923	
$K(\beta_0)$		0.2902	0.2902

Chapter 6

Secondary Storage

1. Introduction

Our first goal is to provide a FCFS (first-come-first-served) queueing analysis sufficiently general to embrace the detailed structure of a large majority of existing systems. The parameters of the mathematical model include a stationary, discrete probability distribution describing the patterns by which requests address information on secondary storage devices, where successive addresses are also allowed to form a first-order Markov chain. Such patterns normally influence system performance, and they are determined by the mechanism that allocates specific storage locations to records (units of information). Thus, in the calculation of conventional performance measures we refer to essentially combinatorial problems in which the effects of different record allocations are studied.

After defining the model, the general analysis of the steady state found in [CofH1] is worked out. We consider only general computational issues; details tend to vary not only with device characteristics but with loading conditions, etc. However, application of the results is illustrated by specializations of the analysis to drum and disk systems. Finally, expressions for conditional waiting times are derived.

In section 3 we then turn to the analysis of the scanning disk introduced in section 1.4. The cylinders addressed are represented by the integers $1,2,...,M$. The read/write head oscillates across the sequence of addresses, visiting cylinders in the order $1,2,...,M-1,M,M-1,M-2,...,2,1,2,...$. The head stops and performs read/write operations at each cylinder where there are waiting requests. The principal objectives of the analysis are queue-length and waiting time distributions. Our presentation is adapted from the work in [CofH2].

Finally, in section 4 we analyze a disk system having two movable read/write heads connected to a single arm-positioning mechanism. We focus on head positioning policies leading to minimum or near-minimum expected head-motion.

The major result is the fixed distance apart that the heads should be maintained in order to minimize expected head motion. Our analysis is patterned after that in [CalCF2].

2. Analysis of FCFS Secondary Storage Devices

We model a secondary storage device as a single-server facility. Incoming requests are Poisson at rate λ. They enter service immediately when the facility is idle. Arrivals at a busy facility wait for service, and there is no limit to the number of such requests that may wait at any given time. All service periods contain an initial period of set-up delay, possibly of zero length. The selection from the queue for service, at the termination of a service period, is done without prior knowledge of the requested service times. During all of our analysis we consider FCFS selection procedures that provide service in the order of arrival.

Requests are of *N types,* which in a specific application will be associated with certain storage unit addresses. The probability that an arriving request is of type j, given that the preceding one was of type i, is p_{ij} and is otherwise independent of the state of the system and its history. These transition probabilities form a matrix $\mathbf{P}$ with an invariant probability vector that we denote by $\mathbf{p} = \mathbf{pP}$. We are interested only in situations where $\mathbf{P}$ is irreducible and all its states are recurrent. The matrix with all rows equal to $\mathbf{p} = (p_1, \ldots, p_N)$ will be denoted by $\overline{\mathbf{P}}$. A request of type j, which is immediately preceded by a request of type i, requests service of duration S_{ij} distributed according to $F_{ij}(\cdot)$, independently of other state descriptors. This service period is generally the sum of two components, $S_{ij} = T_{ij} + K_j$, where T_{ij}, the set-up time, is the time it takes the service facility to switch over from a state of having finished the service for a type j request. The quantity K_j depends on the request type, does not depend on the state of the system, and usually represents the actual transmission time of the information. In most of the intended applications the variables S_{ij} are in fact very simple functions of i and j.

In some situations we must distinguish the service rendered to a request that starts a busy period (i.e., it finds upon arrival an idle system). Invariably, the set-up time T_{ij} is affected; its value under these circumstances is denoted by T_{ij}^0. Associated with T_{ij}^0 is a service duration S_{ij}^0, but the value of K_j is not changed. The arrival process of requests is assumed to be Poisson at rate λ.

We are interested primarily in steady-state behavior. We observe the system at the epochs of departure of requests. Since arrivals and departures happen singly, the distribution of the states of the system at these epochs is the same as at the arrival epochs. Because of the Poisson arrivals these distributions are also equal to the long-term distribution [Coo]. We let X_n denote the number of requests in the system immediately following the departure of the nth request, the one in service included. We let η_n denote the waiting time of the nth request, which terminates at the beginning of the nth set-up time. We let S be the random variable denoting general service time, $F(\cdot)$ the corresponding distribution, and $\tilde{F}(\cdot)$ its Laplace-Stieltjes transform (LST).

Analysis of the Model — The major difference between the model investigated below and standard queueing models is the dependence between successive services. Depending on the type of device and its operating procedures, this relationship may even extend across an intervening idle period. We begin with an analysis that is independent of the order of arrivals. Later, we evaluate the waiting times for a FCFS queue.

First, as usual in queueing systems, the input rates that the facility can sustain must be less than $\lambda_{\max} = 1/E(S)$, where $E(S) = \sum_i \sum_j p_i p_{ij} E(S_{ij})$. This statement is not proved explicitly here, but we note the occurrence of the corresponding discontinuity point in certain of the results.

We observe that $(X_n, J_n;\ n = 1,2,...)$, where J_n is the type of the nth departing request, is an aperiodic, irreducible and, for sufficiently low input rates, recurrent Markov chain. We proceed first to evaluate the generating function of the steady-state distribution of the number in system. This constitutes the heart of the analysis.

We define for $1 \leqslant i \leqslant N$, $k \geqslant 0$, $p_i(k) = \lim_{n \to \infty} Pr[X_n = k | J_n = i]$ and the generating function $G_i(z) = \sum_{k=0}^{\infty} p_i(k) z^k$. The dynamics of our Markov chain are embodied in the matrix $\mathbf{P}$ and the relation $X_{n+1} = X_n - U_n + Y_{n+1}$, where U_n is 0 when $X_n = 0$ and is 1 otherwise, and where Y_{n+1} is the number of arrivals during the service of the $(n+1)$st request. We proceed in a standard way to obtain directly

$$Pr\{X_{n+1}=k|J_{n+1}=j\}Pr\{J_{n+1}=j\} =$$

(1)
$$\sum_{i=1}^{N} Pr\{J_n=i\} \left[Pr\{X_n=0|J_n=i\}Pr\{Y_{n+1}=k,J_{n+1}=j|X_n=0,J_n=i\} \right.$$

$$\left. + \sum_{r=1}^{\infty} Pr\{X_n=r|J_n=i\}Pr\{Y_{n+1}=k-r+1,J_{n+1}=j|X_n=r,J_n=i\} \right]$$

$$k \geqslant 0, \quad 1 \leqslant j \leqslant N$$

The distribution of Y_{n+1} is now derived. It obviously depends on the duration of service of the $(n+1)$st request. As mentioned above, we distinguish between a departure followed by an idle period (with a subsequent service distributed according to $F_{ij}^0(\cdot)$) and a departure for which the next service commences immediately (and is distributed according to $F_{ij}(\cdot)$); the set-up duration may be different in the two cases.

Hence,

(2)
$$Pr\{Y_{n+1}=k|X_n=0, J_n=i, J_{n+1}=j\} = \int_0^{\infty} \frac{(\lambda s)^k}{k!} e^{-\lambda s} F_{ij}^0(ds)$$

and

(3)
$$Pr[Y_{n+1}=k|X_n>0, J_n=i, J_{n+1}=j] = \int_0^{\infty} \frac{(\lambda s)^k}{k!} e^{-\lambda s} F_{ij}(ds)$$

After removing the conditioning on J_{n+1}, we substitute (2) and (3) into (1), multiply by z^k and sum over all k. Since the Markov chain is recurrent, we may drop the subscripts n and $n+1$ to obtain the limiting equation

$$p_j G_j(z) = \sum_{i=1}^{N} p_i p_{ij} \left\{ \pi_i \sum_{k=0}^{\infty} \int_0^{\infty} \frac{(\lambda s z)^k}{k!} e^{-\lambda s} F_{ij}^0(ds) \right.$$

$$\left. + \sum_{r=1}^{\infty} z^r Pr\{X=r|J=i\} \sum_{k=r-1}^{\infty} z^{-1} \int_0^{\infty} \frac{(\lambda s z)^{k-r+1}}{(k-r+1)!} e^{-\lambda s} F_{ij}(ds) \right\}$$

where $\pi_i = Pr\{X=0|J=i\}$. Letting $\tilde{F}_{ij}$ and $\tilde{F}_{ij}^0$ denote the respective transforms of F_{ij} and F_{ij}^0, the above may be rendered as

$$(4) \qquad G_j(z) = \sum_{i=1}^{N} \{\pi_i R_{ij}^0(\lambda(1-z))+z^{-1}[G_i(z)-\pi_i]R_{ij}(\lambda(1-z))\} ,$$

where R_{ij} (respectively R_{ij}^0) is given by $p_i p_{ij} \tilde{F}_{ij}/p_j$ (respectively $p_i p_{ij} \tilde{F}_{ij}^0/p_j$). The various changes of order of summation are allowed since all the sums are trivially absolutely convergent.

The N equations can be written in the more convenient and compact matrix form

$$(5) \qquad\qquad \mathbf{A}(z)\mathbf{G}(z) = \mathbf{B}(z)\boldsymbol{\pi} ,$$

where $\boldsymbol{\pi}$ and $\mathbf{G}(z)$ are the obvious vectors and

$$(6) \qquad\qquad A_{ij}(z) = z\delta_{ij}-R_{ji}(z)$$
$$B_{ij}(z) = zR_{ji}^0(z)-R_{ji}(z)$$

where δ_{ij} is 1 if $i = j$ and is 0 otherwise. Equation (5) has the formal solution

$$(7) \qquad\qquad \mathbf{G}(z) = \mathbf{A}^{-1}(z)\mathbf{B}(z)\boldsymbol{\pi} .$$

The unknown boundary probabilities π_i now have to be determined. First we have

$$(8) \qquad\qquad \mathbf{G}(1) = \mathbf{1} .$$

Second, letting $\mathbf{C}(z)$ be the adjoint matrix of $\mathbf{A}(z)$, and hence $\mathbf{C}(z)\mathbf{A}(z) = |\mathbf{A}(z)|\mathbf{I}$, then we must have

$$(9) \qquad\qquad \mathbf{C}(\xi)\mathbf{B}(\xi)\boldsymbol{\pi} = \mathbf{0}$$

at all points ξ, $|\xi| \leq 1$, that are solutions of

$$(10) \qquad\qquad |\mathbf{A}(z)| = 0 .$$

Using classical results it is not difficult to verify that $|\mathbf{A}(z)|$ vanishes at $z = 1$ and at precisely $N-1$ points that satisfy $|z| < 1$. (See [CofH1].)

Each of the equations in the system (9) is homogeneous, and thus (9) has to be supplemented by an equation that is inhomogeneous. Equation (8) does not give this directly, and we obtain it by noting that if π is the probability an incoming request finds an empty system, then balance equations yield

(11)
$$\pi = \sum_{i-1}^{N} p_i \pi_i \,,$$

and

(12)
$$1-\pi = \lambda \sum_i \sum_j p_i p_{ij} \pi_i [E(S_{ij}^0) - E(S_{ij})] + \lambda \sum_i \sum_j p_i p_{ij} E(S_{ij}) \,.$$

Equations (11) and (12) can now be combined to yield the necessary addition to (9),

(13)
$$\sum_i \pi_i p_i \{1 + \lambda \sum_j p_{ij} [E(S_{ij}^0) - E(S_{ij})]\} = 1 - \lambda E(S) \,,$$

where as before, $E(S) = \sum_i \sum_j p_i p_{ij} E(S_{ij})$. Moments of the number in system can be found in the usual way from the derivatives of $G_i(z)$, $1 \leqslant i \leqslant N$, at $z = 1$.

Before getting into the performance of specific devices, let us consider briefly the problem of calculating waiting times. Let η_n denote the waiting time of the nth request, S_n the duration of its service, and t_n the time between its arrival and that of the $(n+1)$st. As in any single-server FCFS queue, we have the recurrence

$$\eta_{n+1} = \max\{0, \eta_n + S_n - t_n\} \,.$$

Define

$$W_{ij}^n(x) = Pr\{\eta_n \leqslant x | \eta_0, J_0 = i, J_n = j\} \,,$$
$$w_{ij}^n = W_{ij}^n(0) \,,$$

and the transform

$$\tilde{W}_{ij}^n(s) = \int_0^\infty e^{-sx} W_{ij}^n(dx) = w_{ij}^n + \int_{0+}^\infty e^{-sx} W_{ij}^n(dx) \,.$$

Using the dependence structure of η_n and S_n, we obtain after routine manipulations

$$(\lambda - s) \tilde{W}_{ij}^{n+1}(s) = -s w_{ij}^{n+1} + \lambda \sum_{k-1}^{N} \{\tilde{W}_{ij}^n(s) R_{kj}(s) - w_{ij}^n [R_{kj} - R_{kj}^0(s)]\} \,.$$

Passing to the limit $n \to \infty$ and assuming stationarity as before, this becomes in matrix form

$$C_1(s)\tilde{W}(s) = C_2(s)w$$

where

$$C_1(s) = (\lambda-s)I-\lambda R^T(s)$$
$$C_2(s) = -sI+\lambda[(R^0(s))^T-R^T(s)]$$

and where w_j, the limit of w_{ij}^n, is independent of i and equal to π_j. Differentiation of this expression at $s = 0$ yields a system of equations that can be solved to obtain the moments of the waiting times for each request type. The calculations are somewhat involved in general; however, specific applications can be expected to provide important simplifications, as illustrated later.

FCFS Drum-Like Devices — We consider a drum that comprises N logical sectors. The number of tracks (bands) is left unspecified. The time required for the ith sector to pass under the read heads is a constant d_i; thus the set-up time, now called rotational latency, is given by

$$(14) \qquad t_{ij} = \begin{cases} \sum_{k=i+1}^{j-1} d_k, & 1 \leqslant i < j \leqslant N, \\ \\ D-\sum_{k=j}^{i} d_k, & 1 \leqslant j \leqslant i \leqslant N, \end{cases}$$

where $D = \sum_{k=1}^{N} d_k$.

We assume that the physical motion of the drum is the only element that creates delays; i.e., electronic switching times are neglected. A similar device, with a slightly simpler distance structure, is magnetic bubble loop memory [Mit]. In terms of record structure our drum consists of one band of a file drum [FulB]. If the d_i are all equal we have a paging drum [Cof1], to which we return later. We specialize the input process by assuming that the types of successive requests are independent and drawn from a distribution $\{p_i\}$. This is a reasonable assumption for a drum, which is normally a system resource shared by many processes.

An interesting question in the design, and hence in the analysis of such devices, is the dependence of service capacity, or delays, on the pattern of use. For the drum as modeled here, it is well known that when requests are processed continuously, which would be the case when the system is saturated, the average rotational latency, $E(T)$, depends on the distribution $\{p_i\}$ of relative frequencies but not on the manner in which the corresponding records are arranged around the circumference. Indeed, from (14) we have easily

$$t_{ij}+t_{ji} + \begin{cases} D-d_i-d_j, & j \neq i, \\ 2D-2d_i, & j = i. \end{cases}$$

Hence,

$$E(T) - \sum_{i=1}^{N}\sum_{j=1}^{N}p_i p_j t_{ij} - \frac{1}{2}\sum_{i=1}^{N}\sum_{j=1}^{N}(t_{ij}+t_{ji})p_i p_j ,$$

and

(15) $$E(T) - [1+\sum_{i=1}^{N} p_i^2]D/2 - \sum_{i=1}^{N} p_i d_i ,$$

for which the claimed invariance is easily seen.

We now show that this property is not retained when idle periods are taken into account. As before we distinguish between a set-up within a busy period (T) and one that commences a busy period (T^0). $E(T)$ is given by (15). To compute $E(T^0)$ we condition on the following sequence of events. A request for sector j is completed and departs, no other request is waiting, and a request for sector i arrives and finds the head over sector M at a distance X from its termination. Thus, $X+t_{Mi}$ is the set-up delay. Now if τ denotes the time between the departure of the request for the jth sector and the arrival of the new one, then we may write for $0 \leqslant x \leqslant d_m$,

(16) $$Pr\{M=m, X \leqslant x\} - \sum_{r=0}^{\infty} Pr\{rD+t_{jm}+d_m-x \leqslant \tau \leqslant rD+t_{jm}+d_m\} ,$$

where r is the number of complete revolutions the drum made between the departure and arrival. But τ is exponentially distributed with parameter λ. Therefore, we may write for the density corresponding to (16)

(17) $p(m,x) = \dfrac{\lambda e^{-\lambda(t_{jm}+d_m-x)}}{1-e^{-\lambda D}}$, $0 \leqslant x \leqslant d_m$, $1 \leqslant m \leqslant N$,

and hence

$$E(T^0) = \sum_i \sum_j p_i p_j \sum_m \int_{x=0}^{d_m} \lambda(x+t_{mi}) e^{-\lambda(t_{jm}+d_m-x)} dx$$

After integration and rather massive cancellations, one obtains

$$E(T^0) = (1 + \sum_i p_i^2) D/2 - 1/\lambda - \sum_i p_i d_i$$

(18) $$+ D \sum_i \sum_j p_i p_j e^{-\lambda t_{ji}}/(1-e^{-\lambda D})$$

Looking at the last term of (18) we see that $E(T^0)$ depends, as claimed, on the relative arrangement of the records. Note also that the dependence on record arrangement is preserved in paging systems where all d_i are equal.

We do not address the combinatorial problem of finding optimal record arrangements minimizing $E(T^0)$, for such problems fall outside the scope of our analysis. However, the problem is of limited interest. Obviously, when the traffic intensity increases, idle times become rarer, and the relative arrangement is thus *least* important just when capacity is most critical.

So far we have considered the problem of arrangement around the circumference, assuming that the distribution $\{p_i\}$ is given. However, drums have multiple bands, so that the "horizontal" distribution of records on the drum influences $\{p_i\}$, and hence $E(T)$ as well as $E(T^0)$. For the effect on capacity, note that the traffic intensity that the drum handles is determined by (15) and implies the following bound on arrival rates:

$$\lambda \leqslant 1/E(S) = 1/[E(T) + \sum_i p_i d_i]$$

$$= 2/[D(1 + \sum_i p_i^2)] .$$

This result shows the measure of distributions that results in efficient operation of the system. In particular, we have the problem of aggregating records so as to obtain vectors $\{p_i\}$ minimizing $\sum\limits_i p_i^2$ subject to $\sum\limits_i p_i = 1$. This problem is most meaningful within our model when we have a paging drum where N is fixed and all d_i are equal to D/N. In this case, each record (page) can be placed at any sector-band coordinate on the drum. Identifying the total set of records by their access probabilities, $\{q_j\}$, the problem is to partition $\{q_j\}$ into N subsets such that $\sum\limits_i p_i^2$ is minimized, where p_i is the sum of the access probabilities in the ith $(1 \leqslant i \leqslant N)$ block of the partition. Combinatorial analyses of algorithms for this problem have been analyzed in [ChaW,EasW].

Next, we specialize the system of equations in (5). We note first that, since the service time is constant,

$$(19) \qquad R_{ij}(\lambda(1-z)) = p_i \exp[-\lambda(1-z)(t_{ij}+d_j)]$$

The transform of S_{ij}^0 is calculated in a way similar to that producing (18). After some effort we obtain

$$(20) \qquad R_{ij}^0 = z^{-1} p_i \exp\left[-\lambda(1-z)(t_{ij} + d_j)\right]$$
$$\left[1 - \exp(-\lambda z t_{ij}) \frac{1 - \exp(-\lambda(1-z)D)}{1 - \exp(-\lambda D)}\right]$$

Thus, we have from (6)

$$(21) \qquad A_{ij}(z) = z\delta_{ij} - p_j \exp[-\lambda(1-z)(t_{ji}+d_i)]$$

and

$$(22) \qquad B_{ij}(z) = -p_j \exp(-\lambda t_{ji}) \exp(-\lambda(1-z)d_i) \frac{1 - \exp(-\lambda(1-z)D)}{1 - \exp(-\lambda D)} .$$

Now in order to use (9), the roots of $|A(z)| = 0$ are required.

Theorem 1. The determinant of $A(z)$ (in (21)) can be expressed as

$$(23) \qquad |\mathbf{A}(z)| = (b - z^N)/(q-1) ,$$

where

$$b = q \prod_{j=1}^{N} [z - p_j(q-1)], \qquad q = \exp[-\lambda(1-z)D] .$$

Proof. Consider $|A(z)|$ as an Nth degree polynomial expression in the term linear in z, with the exponentials regarded as coefficients. At the N values of z given by $z_j = p_j(q-1)$ (when we treat q as an explicit coefficient of z rather than display its functional dependence), the determinant can be easily evaluated, and we obtain $-z_j^N/(q-1)$. The right-hand side of (23) is a polynomial expression of degree N, which correctly interpolates $|A(z)|$ at the $N+1$ points† $z = z_j$ and $z = 1$, and is therefore the unique interpolating polynomial expression of degree N. ∎

Although we have no intuitive explanation, it is interesting to find that the roots of this equation turn out not to depend to any extent on lengths of individual records (sectors), but merely on their frequency of use. For specific applications the solution of $|A(z)| = 0$ can now be approached by conventional numerical methods.

The Smallest-Latency-First Paging Drum — For the SLF paging drum, we have $d_i = D/N$, $1 \leqslant i \leqslant N$, which corresponds to the constant time to transfer a page. The service mechanism is no longer FIFO among requests of different types (i.e. for pages on different sectors); at any time, the next request served is that one of those waiting whose starting address is nearest the current angular position of the drum. However, waiting requests within a given type create ties in the SLF rule, and these are assumed to be resolved by the FIFO discipline.

To analyze the SLF paging drum we make the simple observation that the processes of serving different request types are mutually independent. We may think of the system as consisting of N parallel FCFS queues with Poisson arrivals at rate $\lambda_k = \lambda p_k$ to the kth queue. Service times, K_k, are constant at D/N for all k. A set-up time while the kth queue is busy is simply $(N-1)D/N$, the time for the drum to rotate past the other $N-1$ sectors. Set-up time, T_k^0, for an arrival to an empty queue is governed by the distribution of the time remaining until the drum has rotated the beginning of sector k back to the service point (read/write heads).

Having found stationary distributions for the N queues in isolation, the system state probabilities are determined from a product distribution. An analysis of this

---

† Obviously $|A(1)| = 0$ since $\Sigma p_j = 1$; a single application of L'Hospital's rule establishes that $z=1$ is a zero of the right-hand side of (23).

model can be found in [Cof1] as a special case of certain results concerning two interacting queues [Ski]. For our purposes, however, these results are more easily obtained by specializing our earlier analysis.

In particular, restricting the analysis leading to (4) to a single sector, say the kth, yields

$$G_k(z) = \pi_k R_{kk}^0(\lambda_k(1-z)) + z^{-1}[G_k(z)-\pi_k]R_{kk}(\lambda_k(1-z))$$

or

(24)
$$G_k(z) = \pi_k \frac{zR_k^0(\lambda_k(1-z))-R_k(\lambda_k(1-z))}{z-R_k(\lambda_k(1-z))},$$

where, for simplicity, $R_k^0 = R_{kk}^0$ and $R_k = R_{kk}$. Now the set-up plus service time, when the queue is non-empty at the last departure, is simply one drum revolution, D. Thus,

(25)
$$R_k(\lambda_k(1-z)) = E[e^{-\lambda_k(1-z)S_k}] = e^{-\lambda_k D(1-z)}.$$

Consider now the set-up time, T_k^0, for the first arrival following a type k departure that left no type k requests behind. As pointed out earlier T_k^0 is the time required for the drum to rotate back to the beginning of sector k. Let r be the number of complete revolutions made by the end of sector k while the queue was empty. Note that $T_k^0 \in [0, \frac{N-1}{N} D]$ if and only if the first arrival occurs during $[rD, rD + \frac{N-1}{N} D]$ for some $r \geqslant 0$, whereas $T_k^0 \in [\frac{N-1}{N} D, D]$ if and only if the first arrival occurs during $[rD + \frac{N-1}{N} D, rD + D]$ for some $r \geqslant 0$. The time delay up to the first arrival is exponentially distributed with parameter λ_k. Thus, the density function, h, for T_k^0 is given by

$$h_k(x) = \begin{cases} \sum\limits_{r=0}^{\infty} \lambda_k e^{-\lambda_k[rD + \frac{N-1}{N} D-x]}, & 0 \leqslant x \leqslant \frac{N-1}{N} D \\[4mm] \sum\limits_{r=0}^{\infty} \lambda_k e^{-\lambda_k[rD + \frac{N-1}{N} D+D-x]}, & \frac{N-1}{N} D < x \leqslant D, \end{cases}$$

or,

$$
(26) \qquad h_k(x) = \begin{cases} \dfrac{\lambda_k}{1-e^{-\lambda_k D}}\, e^{-\lambda_k[\frac{N-1}{N}D-x]}, & 0 \leqslant x \leqslant \dfrac{N-1}{N}D \\[4mm] \dfrac{\lambda_k}{1-e^{-\lambda_k D}}\, e^{-\lambda_k[\frac{2N-1}{N}D-x]}, & \dfrac{N-1}{N}D < x \leqslant D. \end{cases}
$$

Thus,

$$
\tilde{F}_k^0(s) = E(e^{-sT_k^0})
$$

$$
= \frac{\lambda}{1-e^{-\lambda_k D}}\left[\int_0^{\frac{N-1}{N}D} e^{-\lambda\frac{N-1}{N}D}e^{-(s-\lambda_k)x}dx + \int_{\frac{N-1}{N}D}^{D} e^{-\lambda_k\frac{2N-1}{N}D}e^{-(s-\lambda_k)x}dx\right].
$$

Integrating and then substituting $\lambda_k(1-z) = s$, we get

$$
(27) \qquad \tilde{F}_k^0(\lambda_k(1-z)) = \frac{1}{z}\left\{e^{-\lambda_k\frac{N-1}{N}D(1-z)} - e^{-\lambda_k\frac{N-1}{N}D}\,\frac{1-e^{-\lambda_k D(1-z)}}{1-e^{-\lambda_k D}}\right\}.
$$

From $S_k^0 = T_k^0 + D/N$, we get

$$
(28) \qquad R_k^0(\lambda_k(1-z)) = \tilde{F}_k^0(\lambda_k(1-z))e^{-\lambda_k D(1-z)/N}.
$$

Substituting (25) and (28) into (24) gives

$$
(29) \qquad G_k(z) = \pi_k\,\frac{z\tilde{F}_k^0(\lambda(1-z))-e^{-\lambda_k\frac{N-1}{N}D(1-z)}}{z-e^{-\lambda_k D(1-z)}}\,e^{-\lambda_k D(1-z)/N}.
$$

Substituting (27) and evaluating π_k in the usual way from $G(1) = 1$, we obtain the result we have been seeking,

$$
(30) \qquad G_k(z) = \frac{1-\lambda_k D}{\lambda_k D}\,\frac{e^{-\lambda_k D(1-z)}-1}{z-e^{-\lambda_k D(1-z)}}\,e^{-\lambda_k D(1-z)/N},
$$

with

$$
(31) \qquad \pi_k = \frac{1-\lambda_k D}{\lambda_k D}\,\frac{1-e^{-\lambda_k D}}{e^{-\lambda_k\frac{N-1}{N}D}}.
$$

Moments can be found in the usual way. For example, the expected number in the kth queue is given by

$$(32) \qquad E(n_k) = \lambda_k D \frac{N+2}{2N} + \frac{(\lambda_k D)^2}{2(1-\lambda_k D)} \ .$$

To obtain waiting times we identify the stationary state probability for the kth queue, $p_n^{(k)}$, with the probability that n requests arrive during the time a request spends in the system. As before this is justified by the fact that the stationary state probabilities at arrival times and departure times are both equal to $\{p_n^{(k)}\}$ for the continuous-time process. Thus, if η is a random variable having the stationary distribution of time-in-system, then we have

$$p_n(k) = E\left[e^{-\lambda_k \eta} \frac{(\lambda_k \eta)^n}{n!}\right] ,$$

and

$$G_k(z) = E\left[e^{-\lambda_k \eta} \sum_{n=0}^{\infty} \frac{(\lambda_k \eta z)^n}{n!}\right] = E[e^{-\lambda_k \eta(1-z)}] \ .$$

Thus, the transform, $\tilde{W}(s)$, of the time-in-system is given by

$$(33) \qquad \tilde{W}(s) = G_k(1-\lambda_k/s) \ .$$

The first moment, of course, can be obtained from Little's result, $E(w_k) = E(n_k)/\lambda_k$, and (32).

Disk-like devices — We consider now the characteristics of a disk pack (or cartridge), with N cylinders (tracks), and a single arm carrying the read heads. For the purposes of our analysis this is functionally identical to any device whose set-up time T_{ij} is merely a function of $|i-j|$, such as magnetic bubble or shift register storage devices. The following will be in disk terminology.

It is customary to consider the set-up time in disks as composed of two parts: seek time, the duration required for the arm to move between cylinders, and a rotational latency similar to the drum.

As we consider here a primitive request-queue management technique, we also limit all explicit calculations concerning disks in two ways: Rotational latency is eliminated by the method of reading, which is to transmit one whole track per request (the portion of the disk passing under a read head during one full revolution). The desired record is subsequently located in memory and perhaps pieced together from two portions. The latter situation occurs when the requested record was under the read head when the seek terminated and transmission started.

In these devices (in contrast to the scanning disks analyzed in the next section) the arm does not react "on the fly" to changes of destination, but rather maintains a "busy" status until a desired seek is terminated and the arm is stopped; only then can a new seek be initiated. Comprehensive discussions of these delays can be found in [Frank] and [Wat].

Although the set-up times T_{ij} of bi-directional tapes conform with the above characterization, we exclude them from this discussion on both practical and analytical grounds. First, rotational latency is of course absent here; also, the tape system can usually handle changes of destination "on the fly" in a much simpler way than in a disk system. (Thus, with sufficiently high processor speeds FCFS may be a less natural operating technique.) Second, analytically we find the dependence structure between successive services even more involved than in our model: T_{ij} depends on the boundary of record i where its reading terminated, and this, in turn, involves the even earlier record. We note, though, that if the idle-period policy were of the type denoted by (a) in the following, one randomization on the identity of that preceding record is enough to properly define the necessary variables.

Unlike the drum, the behavior of the system when no requests are pending may have different modes. The more common ones are (in disk terminology)

(a)   The arm remains in place, at the cylinder last used.

(b)   the arm is reset at a fixed position, say cylinder r.

These modes determine the distributions of the respective S_{ij}^0. In case (a) it is clear that S_{ij}^0 and S_{ij} have the same distribution. In case (b) let $f(i,j)$ be the time taken to travel from cylinder i to cylinder j when no intervening cylinders are read. Normally, this is the same as the set-up time T_{ij} and we assume so in the following. The set-up time T_{ij}^0 succeeding an idle period is then given by

$$T^0_{ij|s} = \begin{cases} f(r,j) & s \geqslant f(i,r) \\ \\ f(i,r)-s+f(r,j) & s < f(i,r) \end{cases}$$

where s is the length of the idle period. Since s is exponentially distributed, we immediately find

$$E(T^0_{ij|s}) = f(i,r)+f(r,j)-[1-\exp(-\lambda f(i,r))]/\lambda .$$

The expected duration of this delay is calculated as follows. Note first that $p^0_i = Pr$ [cylinder i was just read|an idle period just started] $= \pi_i p_i / \sum_k \pi_k p_k$ where π_i, as defined earlier, is the probability that the request queue is empty following the completion of service from cylinder i. Thus we obtain

$$E(T^0) = \sum_i \sum_j p^0_i p_j E_s(T^0_{ij})$$

and

$$E(T^0) = \sum_j p_j f(r,j) + \sum_i \pi_i p_i f(i,r)\exp(-\lambda f(i,r))/(\lambda \sum_k \pi_k p_k) - 1/\lambda$$

The value of $E(T^0)$ does not influence the overall service capacity of the system. It is a factor in its response when not fully loaded, and becomes more important as the load becomes lighter.

Unlike the drum, which is a constant speed system, we have here important acceleration and deceleration effects. An approximation that holds for a rather large subset of available disks is

$$f(i,j) = \begin{cases} 0, & i = j \\ \\ A+B|i-j|, & i \neq j \end{cases}$$

where A incorporates the effects of the changes of speed of motion of the arm and B corresponds to movement at constant speed. The approximation is not very good for short distances and quite acceptable when a sizeable portion of the disk radius has to be traversed. This completes the specification, so that the earlier analysis can be applied. (Nothing comparable to Theorem 1 has been found here, however.)

3. Analysis of the Scanning Disk

The scanning-disk model developed below specializes a model [Eis] in which multiple queues receive periodic service in fixed but arbitrary cycles. In particular, the disk is viewed as comprising M service points (cylinders) arranged along a line. At any point in time the server (arm) is either located at one of these points (possibly performing an I/O operation) or it is in motion between them (seeking). In this model seeks are done to adjacent cylinders only and require a time units. Each I/O operation requires a constant, fixed amount of time, T, to complete.[†] As usual we make no distinction between reads and writes. Request arrivals for the m-th cylinder constitute a Poisson process with rate λ_m. The arrival processes to distinct cylinders are assumed to be independent of each other and the state of the system. The order of service at each cylinder is FCFS. Transition times between seek termination and beginning of service, as well as between services are assumed to be zero. Our treatment of this model follows closely that given in [CofH2].

Elements of the Model — We describe arm motion in terms of *stages*. The arm moves in cycles of stages numbered 1 to $2M-2$; the correspondence between the i-th stage and cylinder m_i is given by

$$(34) \qquad m_i = \begin{cases} i & 1 \leqslant i < M \\ 2M-i, & M \leqslant i \leqslant 2M-2 \end{cases}$$

Note that only one stage corresponds to each of cylinders 1 and M. Otherwise, stages i and $2M-i$ correspond to the servicing of requests at cylinder m_i, and a cycle through the $2M-2$ stages corresponds to a complete scan of the arm across the disk and back.

The main process we investigate is $\mathbf{N}(t)$, the vector of queue lengths at time t. The state of the disk and the collection of queues is observed at instants when a

---

† We see later that this restriction can be removed.

specification of the value of $N(t)$ and the position of the arm gives a complete (i.e. Markovian) description of the disk facility. A value of $N(t)$ is denoted by an M-vector $\mathbf{n} = (n_1,...,n_M)$. We observe the state of the system at stage terminations and define $\beta_{\mathbf{n}}^i$ as the probability that immediately after a stage-i termination the queues are in state $\mathbf{n}$. Clearly, since no requests for cylinder m_i are outstanding at a stage-i termination, $\beta_{\mathbf{n}}^i > 0$ implies $n_{m_i} = 0$. The states at stage terminations describe an irreducible aperiodic Markov chain.

We state without proof the intuitive claim that the chain is recurrent when $\lambda = \sum_{m=1}^{M} \lambda_m < 1/T$, which is the same as requiring $\Sigma \rho_m < 1$, where $\rho_m = \lambda_m T$ is the traffic intensity of queue m. Consequently, the states of the system at stage terminations may be assumed to have the stationary probabilities $\beta_{\mathbf{n}}^i$. Note that the recurrence condition depends only on λ and T, and not on a. Indeed, the fraction of time the arm spends seeking vanishes in the limit $\Sigma \rho_m \to 1$. At the end of this section we comment further on this point.

In order to calculate state probabilities we find it expedient to consider an additional set of regeneration points, viz. stage beginnings, and define $\alpha_{\mathbf{n}}^i$ as the probability that immediately before a stage-i beginning the system is in state $\mathbf{n}$. As before, the computational tools are generating functions

$$(35) \qquad \beta^i(\mathbf{z}) = \sum_{n_1=0}^{\infty} \cdots \sum_{n_m=0}^{\infty} \beta_{\mathbf{n}}^i z_1^{n_1}...z_M^{n_M}$$

with $\alpha^i(\mathbf{z})$ similarly defined. Note that in (35) the sum over n_{m_i} contributes only when $n_{m_i} = 0$ by virtue of its being a stage-i termination state, and thus could be suppressed.

The distribution function (df) of the numbers of arrivals $(n_1,n_2,...,n_M)$ to the M cylinders during a period with df $F(\cdot)$ is given by

$$(36) \qquad p(F;n_1,...,n_M) = \int_0^{\infty} \frac{(\lambda_1 t)^{n_1}}{n_1!} \cdots \frac{(\lambda_M t)^{n_M}}{n_M!} e^{-\lambda t} F(dt) .$$

If $\tilde{F}(\cdot)$ is the transform of $F(\cdot)$, then the generating function $p(\cdots)$ is given by

$$(37) \qquad p(F;\mathbf{z}) = \tilde{F}(\lambda_1-\lambda_1 z_1+...+\lambda_M-\lambda_M z_M) \equiv \hat{F}(\mathbf{z}) .$$

Note that $\hat{F}(\cdot)$ has a vector argument.

Let $A(s)$ and $C(s)$ denote the transforms of the seek and service periods, respectively, with $\hat{A}(z)$ and $\hat{C}(z)$ defined in analogy with $\hat{F}(z)$. Since these periods are in fact constants we have

$$(38) \qquad \hat{A}(z) = e^{-a\sum_m \lambda_m (1-z_m)}, \quad \hat{C}(z) = e^{-T\sum_m \lambda_m (1-z_m)} .$$

Calculation of $\beta^i(z)$ — Since the queues at stage beginnings contain just those requests that were there at the last stage termination plus those that arrived during the seek, and since these two components are independent, we have

$$(39) \qquad \begin{aligned} \alpha^i(z) &= \beta^{i-1}(z)\hat{A}(z), \quad i > 1 \\ &= \beta^{2M-2}(z)\hat{A}(z), \quad i = 1 \end{aligned}$$

thus relating states "across" a seek.

To obtain an equation satisfied by the $\beta(\cdot)$ alone we need to relate the states of the system "across" a complete servicing of a stage. Let stage-i of queue m_i start with k_{m_i} requests in the queue. Restricting our attention to this queue until it empties, the analysis of a standard M/G/1 system applies. Denote by $G_{m_i}(\cdot)$ the df of a busy period in such a queue and by $\tilde{G}_{m_i}(\cdot)$ its transform. Then $\tilde{G}_{m_i}(\cdot)$ satisfies the equation

$$(40) \qquad \tilde{G}_{m_i}(s) = C(s+\lambda_{m_i}-\lambda_{m_i}\tilde{G}_{m_i}(s)) .$$

We use this relation later.

A compound busy period, which begins with k_{m_i} customers present, is the sum of k_{m_i} i.i.d. simple busy periods. Thus it has the df $G_{m_i}^{*^{k_{m_i}}}(\cdot)$. By observing queue lengths when this stage terminates we find for the number, h_i, of requests in queue m at a stage-i termination

$$(41) \qquad h_i = \begin{cases} 0, & m = m_i \\ \\ k_m + c(\lambda_{m_i}; G_{m_i}^{*^{k_{m_i}}}), & m \neq m_i . \end{cases}$$

On the right-hand side, for $m \neq m_i$, k_m is the number of customers at queue m when stage-i started, and is distributed according to $\alpha_{\mathbf{n}}^i$; and $c(\lambda;F)$ is a random variable distributed as the number of events occurring in a homogeneous Poisson process with rate λ during a period with the df $F(\cdot)$. Adapting (36) we may write $p_{m_i}^{(i)}(G_{m_i}^{\bullet^{k_m}};\cdot)$ for the joint df of this latter variable over $M-1$ queues, excluding m_i; the superscript signifies the suppression. Thus,

$$(42) \qquad \beta_{\mathbf{n}}^i = \sum_{\mathbf{k} \leqslant \mathbf{n}} \alpha_{\mathbf{k}}^i p_{m_i}^{(i)}(G_{m_i}^{\bullet^{k_m}}; \mathbf{n-k}) \;,$$

where the sum is over all $\mathbf{k}$ such that $\mathbf{n} \geqslant \mathbf{k}$ component-wise (excepting the m_i-th component), and $\mathbf{n} = \mathbf{k}$ when $k_{m_i} = 0$. Substituting for $p^{(i)}(\cdot\cdot)$ yields

$$\beta_{\mathbf{n}}^i = \sum_{\mathbf{k} \leqslant \mathbf{n}} \alpha_{\mathbf{k}}^i \int_0^\infty e^{-(\lambda-\lambda_{m_i})t} \prod_{m \neq m_i} \frac{(\lambda_m t)^{n_m-k_m}}{(n_m-k_m)!} \, G_{m_i}^{\bullet^{k_m}}(dt) \;,$$

where $\mathbf{k} \leqslant \mathbf{n}$ should be construed as above. Multiplying by $\prod_{r \neq m_i} z_r^{n_r}$ on both sides we obtain, summing on all $\mathbf{n}$,

$$\beta^i(\mathbf{z}) = \sum_{\mathbf{n}} \prod_{r \neq m_i} z_r^{n_r} \beta_{\mathbf{n}}^i$$

$$= \sum_{\mathbf{k}} \alpha_{\mathbf{k}}^i \prod_{r \neq m_i} z_r^{k_r} \int_0^\infty e^{-(\lambda-\lambda_{m_i})t} \sum_{\mathbf{n} \geqslant \mathbf{k}} \prod_{m \neq m_i} \frac{(\lambda_m z_m t)^{n_m-k_m}}{(n_m-k_m)!} \, G_{m_i}^{\bullet^{k_m}}(dt)$$

$$= \sum_{\mathbf{k}} \alpha_{\mathbf{k}}^i \prod_{r \neq m_i} z_r^{k_r} \int_0^\infty e^{-(\lambda-\lambda_{m_i})t + \sum_{m \neq m_i} \lambda_m z_m t} \, G_{m_i}^{\bullet^{k_m}}(dt)$$

Performing the integration and noting that $\lambda-\lambda_{m_i} = \sum_{m \neq m_i} \lambda_m$, one gets

$$\beta^i(\mathbf{z}) = \sum_{\mathbf{k}} \alpha_{\mathbf{k}}^i \prod_{r \neq m_i} z_r^{k_r} \tilde{G}_{m_i}^{k_i} \left[\sum_{m \neq m_i} \lambda_m (1-z_m) \right] \;,$$

and summing over $\mathbf{k}$ we obtain

$$(43) \qquad \beta^i(\mathbf{z}) = \alpha^i(\mathbf{z}^{(i)}) \;,$$

where the superscript i over z means that the m_i-th component is replaced by $\tilde{G}_{m_i}\left[\sum_{k \neq m_i} \lambda_k - \lambda_k z_k\right]$. Combining (39) with (43) we finally obtain

$$\text{(44)} \qquad \beta^i(z) = \beta^{i-1}(z^{(i)})\hat{A}(z^{(i)})$$

which is the basis for the numerical calculations developed later. This equation could also be used, as in [Ei], for the basis of a formal solution for the β functions, but as this solution is of limited utility for us, it is not presented here.

Refining the Chains — The chain embedded at stage terminations gives a description of the evolution of $N(t)$ that is too "coarse" to define the waiting times of individual requests. To evaluate these we imbed in $N(t)$ a finer chain defined at service beginning and completion epochs. These epochs are regeneration points for $N(t)$, and thus its values there also constitute an ergodic Markov chain. Let $\pi_{i,\mathbf{n}}$ be the probability that a service termination occurs in stage-i, and the value of N just after the termination is $\mathbf{n}$.† Following the relation between $\beta_\mathbf{n}^i$ and $\alpha_\mathbf{n}^i$, we may relate $\pi_{i,\mathbf{n}}$ to the joint probability, $\omega_{i,\mathbf{n}}$, of observing the system just before a service initiation in stage-i and at state $\mathbf{n}$. A relation similar to (39) between the generating functions of $\pi_{i,\mathbf{n}}$ and $\omega_{i,\mathbf{n}}$ is simple, since they are related across a single service duration. Hence

$$\text{(45)} \qquad \pi_i(z) = \omega_i(z)\hat{C}(z)/z_{m_i}$$

The remainder of this section is devoted to finding a relation between the $\pi_{i,\mathbf{n}}$ and $\beta_\mathbf{n}^i$. The analytical development in [Ei] will be replaced here by one that exploits somewhat more intuitive arguments.

We focus on the ith stage and consider observations made just after completions of stage-i services and seeks from stage $i-1$ to stage i, i.e. the union of those epochs that define transitions of the chains described by $\pi_{i,\mathbf{n}}$ and $\alpha_\mathbf{n}^i$. Next, suppose there have been K requests served by the system. For large K, the

---

† The difference in notation between $\beta_\mathbf{n}^i$ and $\pi_{i,\mathbf{n}}$ reflects a difference in definition; in $\beta_\mathbf{n}^i$ i specifies a conditioning event, but in $\pi_{i,\mathbf{n}}$, i is a part of the state descriptor.

number of epochs at which state $\mathbf{n}$ occurred is approximately $\alpha^i(K)\alpha_{\mathbf{n}}^i + K\pi_{i,\mathbf{n}}$, where $\alpha^i(K)$ was the number of stage-i beginnings.

Now consider the epochs just after the *beginnings* of stage-i services and seeks from stage i to stage $i+1$. These have a one-to-one correspondence with the set of epochs defined above, and correspond also to the union of the set of epochs of transition of the chains described by $\omega_{i,\mathbf{n}}$ and $\beta_{\mathbf{n}}^i$. But from the point of view of these latter epochs we have $\beta^i(K)\beta_{\mathbf{n}}^i + K\omega_{i,\mathbf{n}}$ as the expected number of epochs at which state $\mathbf{n}$ occurred, where $\beta^i(K)$ was the number of stage-i completions. Thus, dividing by K and taking the limit $K \to \infty$ we have by the law of large numbers

$$(46) \qquad \gamma\beta_{\mathbf{n}}^i + \omega_{i,\mathbf{n}} = \gamma\alpha_{\mathbf{n}}^i + \pi_{i,\mathbf{n}}$$

where $\gamma = \lim_{K \to \infty} \beta^i(K)/K = \lim_{K \to \infty} \alpha^i(K)/K$ is the limiting ratio of the number of stage-i visits to the number of requests served, and must be the same for all i, since each stage occurs once per cycle. In terms of generating functions

$$(47) \qquad \gamma\beta^i(z) + \omega_i(z) = \gamma\alpha^i(z) + \pi_i(z)$$

Substituting for $\omega_i(z)$ from (45) to obtain

$$(48) \qquad \pi_i(z) = \gamma C(z)\, \frac{\alpha^i(z) - \beta^i(z)}{z_{m_i} - C(z)}$$

which, in conjunction with (39), gives us the desired relation between the $\pi_i(z)$ and $\beta^i(z)$. To evaluate γ we may proceed by the following brief expected-value argument.

In equilibrium, the average number served per cycle is given by λD where D is the average cycle length. Since for each i there is exactly one stage-i visit per cycle, we have $\gamma = 1/\lambda D$. Since the total seek time per cycle is $2(M-1)a$, we have for the average cycle time, $D = \lambda DT + 2(M-1)a$. Hence,

$$D = \frac{2(M-1)a}{1 - \lambda T}$$

and

$$(49) \qquad \gamma = \frac{1 - \lambda T}{2(M-1)\lambda a}$$

The linear dependence of the average cycle duration D on the seek time a may seem strange when the limit $a \to 0$ is considered, as it would predict the vanishing of D regardless of ρ. However, this merely represents the ability of the arm to make, in the limit $a \to 0$, an unbounded number of cycles during each "idle period" (idle in the sense that no requests are available); the time between idle periods has a finite mean, so D would be "biased away." The unboundedness of γ is then clear.

This effect also makes the term "utilization" a rather inappropriate term for $\rho = \lambda T$. In queueing models one associates high or low values of ρ with sluggish or prompt response. Here, λT merely denotes the fraction of the cycle time used for actual transmission. The responsiveness of the device depends upon the cycle *duration* and hence on a. For low values of λ the mean response time is essentially determined by a.

Calculation of Waiting Times — Let W_m and W^i denote respectively the waiting times of a request at queue m and one that is served during stage-i. Consider for the moment one queue in isolation. We observe that since the service order is FIFO, the requests queued at service termination must have arrived during the waiting or service time of the request just completed. Now let

$$\pi_{i,j} = \sum_{k \in \{n | n_{m_i} = j\}} \pi_{i,k}$$

be the marginal queue-length distribution of queue m_i at stage-i service completions. Define $\pi_i = \sum_n \pi_{i,n} = \sum_j \pi_{i,j} = \pi_i(1)$. Then from the above observation we have for the probability that queue m_i holds $n_{m_i} = j$ customers at service completion epochs

(50)
$$\pi_{i,j}/\pi_i = \int_0^\infty \frac{(\lambda_{m_i} t)^j}{j!} e^{-\lambda_{m_i} t} F_{W^i} * F_T(dt) \ ,$$

where $F_{W^i} * F_T(t)$ corresponds to the convolution of the distributions for stage-i waiting times and service times. $F_T(\cdot)$ is the distribution of a constant T; therefore, calculating the generating functions of both sides of (50) we obtain for the transform of F_{W^i}

$$(51) \qquad \tilde{F}_{W^i}(s) = \frac{\pi_i\left(1,...,1 - \dfrac{s}{\lambda_{m_i}},...,1\right)}{\pi_i \exp(-Ts)}$$

Denoting the generating function of $\pi_{i,j}$ by $\pi_i(z)$, the numerator of (51) is $\pi_i(1 - s/\lambda_{m_i})$. For the distribution $F_{W_m}(\cdot)$ of waiting time at cylinder m, we average over the stages where $m_i = m$ and get

$$F_{W_1}(t) = F_{W^1}(t) ,$$

$$(52) \qquad F_{W_m}(t) = \frac{\pi_m F_{W^m}(t) + \pi_{2M-m} F_{W^{2M-2m}}(t)}{\pi_m + \pi_{2M-m}}, \qquad 1 < m \leqslant M .$$

Next, from (39) and (48) we can express $\pi_i = \pi_i(1)$ in terms of the first derivatives of $\beta^{i-1}(z)$ at $z = 1$,

$$(53) \qquad \pi_i = \frac{\gamma}{1 - \rho_{m_i}} \left(\beta_{m_i}^{i-1} + a\lambda_{m_i}\right) ,$$

where $\beta_m^i \equiv \dfrac{\partial \beta^i(z)}{\partial z_m}\big|_{z=1}$ and $\rho_{m_i} = \lambda_{m_i} T$. Finally, we differentiate (51) at $s = 0$ using (48) and then again (39) to eliminate $\alpha^i(\cdot)$. In the resulting expression we substitute for γ and π_i from (49) and (53) and obtain

$$(54) \qquad E(W^i) = \frac{\beta_{m_i m_i}^{i-1} + 2a\lambda_{m_i}\beta_{m_i}^{i-1} + a^2\lambda_{m_i}^2}{2\lambda_{m_i}(a\lambda_{m_i} + \beta_{m_i}^{i-1})} + \frac{T\rho_{m_i}}{2(1 - \rho_{m_i})} ,$$

where

$$\beta_{m,n}^i \equiv \frac{\partial^2 \beta^i(z)}{\partial z_m \partial z_n}\big|_{z=1} .$$

From (52) the desired mean waiting time is

$$E(W_1) = E(W^1) ,$$

$$(55) \qquad E(W_m) = \frac{\pi_m E(W^m) + \pi_{2M-m} E(W^{2M-m})}{\pi_m + \pi_{2M-m}}, \qquad 1 < m \leqslant M .$$

The lengths, N_m, of queues accumulated at the individual cylinders are of practical interest as well. Using Little's theorem, the mean value, $E(N_m)$, at request completion epochs is given by $\lambda_m E(W_m)$. To obtain higher moments one needs only to differentiate (48) further, and accumulate contributions as in (52). However, the complexity of these calculations can be expected to increase significantly.

Numerical Calculations — The expressions for $E(W^i)$ and higher moments include the partial derivatives of the functions $\beta^i(z)$ at $z = 1$. We show here how these can be calculated based on (44). We also use the following values and notation:

(56)
$$-\tilde{G}'_m(0) = T/(1-\rho_m) \equiv T\gamma_m \quad \text{(see (40))} ,$$

$$\tilde{G}''_m(0) = T^2\gamma_m^3 ,$$

$$\zeta_m^{m_i}(\bar{z}) \equiv \frac{\partial z_{m_i}^{(i)}}{\partial z_m} = -(1-\delta_{m,m_i})\lambda_m \tilde{G}'_{m_i}\left(\sum_{j\neq m_i}\lambda_j(1-z_j)\right) ,$$

(57)
$$\zeta_m^{m_i} \equiv \zeta_m^{m_i}(1) = (1-\delta_{m,m_i})\rho_m\gamma_{m_i} .$$

We get by straightforward calculation

$$\frac{\partial \hat{A}(z^{(i)})}{\partial z_m} = a(1-\delta_{m,m_i})\hat{A}(z^{(i)})[\lambda_m + \lambda_{m_i}\zeta_m^{m_i}(z)] .$$

At $z = 1$ the right-hand side becomes $(1-\delta_{m,m_i})a\lambda_m\lambda_{m_i}$.

Proceeding from (44) we get

$$\beta_m^i(z) \equiv \frac{\partial \beta^i(z)}{\partial z_m} = (1-\delta_{m,m_i})\{\beta_m^{i-1}(z^{(i)})\hat{A}(z^{(i)})$$

(58)
$$+ \beta_{m_i}^{i-1}(z^{(i)})\hat{A}(z^{(i)})\zeta_m^{m_i}(z) + a\beta^i(z)(\lambda_m+\lambda_{m_i}\zeta_m^{m_i}(z))\} .$$

At $z = 1$ these derivatives yield

(59)
$$\beta_m^i = (1-\delta_{m,m_i})\{\beta_m^{i-1}+\beta_{m_i}^{i-1}\rho_m\gamma_{m_i}+a\lambda_m\gamma_{m_i}\} \quad 1 \leqslant i \leqslant 2M-2, \; 1 \leqslant m \leqslant M .$$

Differentiating $\beta_m^i(z)$ with respect to z_n, and evaluating at $z = 1$, we find

(60)
$$\beta_{m,n}^i = (1-\delta_{m,m_i})(1-\delta_{n,m_i})\{\beta_{m,n}^{i-1}+\beta_{m,m_i}^{i-1}\zeta_n^{m_i}+\beta_{n,m_i}^{i-1}\zeta_m^{m_i}$$
$$+ \beta_{m_i,m_i}^{i-1}\zeta_m^{m_i}\zeta_n^{m_i}+\beta_m^{i-1}a\lambda_n\gamma_{m_i}+\beta_{m_i}^{i-1}\zeta_m^{m_i}a\lambda_n\gamma_{m_i}$$
$$+ \beta_{m_i}^{i-1}\zeta_m^{m_i}\zeta_n^{m_i}\gamma_{m_i}+\beta_n^i a\lambda_m\gamma_{m_i}+a\lambda_{m_i}\zeta_m^{m_i}\zeta_n^{m_i}\gamma_{m_i}\} .$$

Higher derivatives are readily formed in this manner.

Equation (59) can be used to calculate all β_m^i. These equations, while quite cumbersome for symbolic manipulation, are very well suited to numerical solution by computer. Equation (59) is applied $2M-2$ times (each time for all values of m) in order to collect coefficients for a set of M equations, linear in (say) β_m^1. These are straightforward to solve; re-applying (59) yields the first derivatives that we need.

A very similar procedure for (60) is used to determine values for the second order derivatives. We note in passing that although the calculation of the mean waiting time required the values of only a small fraction of these derivatives, the form of the only expression we found that was relatively convenient to solve required the evaluation of many more in the process.

The regularity of the system of equations (60) suggests that the corresponding matrices might be explicitly inverted. This does not seem to be the case: They can be inverted in terms of suitably defined $M \times M$ blocks, but inverting these blocks results in an overall effort of the same complexity.

Since $\beta_{m_i}^{i-1}$, as required in (54), has the value of the expected queue size at queue m_i when stage $i-1$ is terminated, the $\beta_{m_i}^i$ may be determined in the following more direct way. Define

> g_i =expected time between the end of stage i and the last time at which the head departed from queue m_i, and

> d_i =expected time between the end of stage $i-1$ and the end of stage i.

We state without proof the following mean-value relationships (they should be obvious on inspection):

$$\beta_{m_i}^{i+1} = \lambda_{m_i} d_{i+1} ,$$
$$d_i = \lambda_{m_i} g_i T + a = \rho_{m_i} g_i + a ,$$
$$g_1 = g_M = D - \sum_{i=1}^{2M-2} d_i ,$$
$$g_2 = d_1 + d_2 = (d_1 + a)/(1 - \rho_{m_2}) ,$$
$$g_i = g_{i-1} + d_i + d_{2M-i+1}, \quad 3 \leqslant i \leqslant M-1 ,$$
$$g_i = D - g_{2M-i} , \quad 2 \leqslant i \leqslant M-1 ,$$

$$g_i = \frac{1}{1-\rho_{m_i}} \left[\frac{\beta_{m_i}^{i-1}}{\lambda_{m_i}} + a \right], \quad 1 \leqslant i \leqslant 2M-2 .$$

The last equation results first from breaking g_i down into the time interval beginning with the departure of the arm from queue m_i and ending with the termination of stage $i-1$, with mean length $\beta_{m_i}^{i-1}/\lambda_{m_i}$, plus the time interval during which queue m_i served. Next, we observe that if the load at queue m_i is A when its processing begins, the arm's expected stay there is $A/(1-\rho_{m_i})$. Finally, then

$$\beta_{m_2}^1 = \lambda_{m_2} d_1, \quad \beta_{m_i}^{i-1} = \lambda_{m_i}(d_{2M-i+1}+g_{i-1}), \quad 3 \leqslant i \leqslant M-1 .$$

Thus, the $\beta_{m_i}^{i-1}$ may be obtained recursively, beginning with (49). Unfortunately, no such procedure was discovered for the second derivatives.

The calculations outlined above were performed in [CofH2] for a number of input rates and distributions (across the disk). The following characterizes the results.

— The total input rate (λ) and cylinder number (m) are the main determinants of $E(W_m)$. It is quite robust under changes in the distribution of λ_m.

— The mean was also only very slightly influenced by relative changes in λ and T, so long as ρ was not affected.

— "Popular" cylinders require shorter waiting times than those with low λ_m.

— When all λ_m are equal the graph of $E(W_m)$ vs. m is parabolic. This is not true for any non-uniform distribution.

Complexity of the Calculations — Each application of (59) uses $O(M^2)$ operations (the number of substitutions per line of the equations increasing from 3 to M), for a total of $O(M^3)$. This is also the time complexity of the solution of the resultant set of $M-1$ equations and the back substitution through (59); the total is then still $O(M^3)$.

Similarly, each application of (60) requires $O(M^3)$ operations; here the solution of the final set of equations dominates, with $O(M^6)$ operations required. Thus, the numerical evaluation of these equations for realistic devices, where M assumes values in the low hundreds is too expensive for most purposes. One should use it, however, to investigate qualitative features, such as

— Dependence of performance measures on hardware parameters, and on details of the algorithms used for scheduling, as well as for comparison with other scheduling methods on the same hardware.

— Evaluation of the quality of various approximations to the above analysis before their applications to a full-fledged system.

Elaborations of the Model — The model as presented captures a number of the properties of a disk system which are critical to its performance. For the sake of simplicity, however, it does contain a good many simplifying assumptions. We discuss here a number of these assumptions and what their removal or modification entails, mainly in terms of model tractability.

1. *Request Service Duration* — This quantity was assumed constant and equal for all cylinders, simply to keep down the level of detail of the model (and use somewhat less storage during numerical calculations). Since these properties are not used explicitly in the procedures we developed, there is absolutely no effect on the analysis when we assume that the time to service a request has a general distribution function, which may indeed be cylinder-specific. Let S_m be the variable representing the service duration at cylinder m, with distribution function and transform, $F_{S_m}(\cdot)$ and $\tilde{F}_m(\cdot)$, respectively; the transform of the busy period distribution is then the solution of the equation $\tilde{G}_m(s) = \tilde{F}_m(s + \lambda_m - \lambda_m \tilde{G}_m(s))$. $\tilde{F}_m(\cdot)$ itself should also be substituted for $\exp(-Ts)$ in the appropriate equations. No essential change need be made, and the calculations proceed in precisely the same manner.

2. *Meaningful Seeks* — By this we refer to the incorporation into the model of the following modification: At stage completion a seek is initiated to the next non-empty cylinder queue in the cycle, as given in (34). We recognize here the fact that, due to acceleration effects the time to traverse k cylinders in a single seek is appreciably less than k times the duration of a single-cylinder seek. This certainly represents better the way a disk facility is managed, but it also introduces a number of complicating factors that will require a new approach:

— Instead of a single seek time, we may now have as many as $M(M-1)$, although their number can usually be reduced to $M-1$ by assumptions that are quite borne out in practice.

— The movements of the arm are not prescribed as before, but rather depend on the state of the system, a feature that is likely to defeat a conventional queueing-theoretic approach.

3. *Dependent Arrivals* — We refer to the addressing patterns of section 2. That is, to the description of the state of the system a further index is added — the identity of the last cylinder addressed by the arrival process. This is a major departure from the model analyzed above, and the methods used there would be ineffective in handling it.

4. An Analysis of a Disk System with Two Read/Write Heads

Consider the mathematical model introduced in chapter 1 for computer disk storage devices having two movable read/write heads [CalCF2]. Storage addresses are approximated by points in the continuous interval [0,1], and requests for information on the disk are processed in a strict first-come-first-served order. We assume that the disk heads are maintained a fixed distance, d, apart, i.e. in processing a request both heads are moved the same distance in the same direction. Both heads must always remain on the disk surface (i.e. in the unit interval); thus, requests in [0,d] and [1-d,1] must be processed by the left and right head, respectively.

Assuming that successive requested locations are independently and uniformly distributed over [0, 1], we calculate the invariant measure of a Markov chain representing successive head positions under the *Nearer-Server* (NS) rule: requests in [0, d] are processed by the left head, those in [1−d, 1] by the right head, and those in [d, 1−d] by the nearer of the two heads. Our major objective is the equilibrium expected distance, E_d, that the heads are moved in processing a request. For the problem of designing the separation distance, d, we show that $\min_d E_d = 0.16059$ at $d = 0.44657$.

Formal definitions are presented next along with preliminary results. Then we calculate expected head motion and find that value of d for which it is minimum.

For purposes of this optimization only a bound on the expected head motion for $d < 1/3$ need be calculated. We conclude with remarks on the significance of the results.

Preliminary Results — Let $X_i \in [0,1-d]$, $i \geqslant 1$, be the position of the left head after the first i requests have been serviced by the NS rule. As noted in [CalCF2] it is easily verified that $\{X_i\}$ is an ergodic Markov chain. The invariant measure of $\{X_i\}$ is denoted by $p(x)dx \equiv p(x,d)dx$, and $\tau(t,x)dx$ denotes the transition probability. We have by definition

$$(61) \qquad p(x) = \int_0^{1-d} \tau(t,x)p(t)dt \, ,$$

and a corresponding expected head motion

$$(62) \qquad E_d = \lim_{i \to \infty} E(|X_{i+1}-X_i|) = \int_0^{1-d} \left[\int_0^{1-d} |x-t|\tau(x,t)dt \right] p(x)dx \, .$$

In the remainder of this subsection we calculate $p(x)$ for $\frac{1}{3} \leqslant d \leqslant \frac{1}{2}$. In the next subsection $\min_{\frac{1}{3} \leqslant d \leqslant \frac{1}{2}} E_d$ is evaluated and it is verified numerically that

$$\min_{0 \leqslant d \leqslant \frac{1}{2}} E_d = \min_{\frac{1}{3} \leqslant d \leqslant \frac{1}{2}} E_d.$$

To work out the functional equation in (61) we may use the function $\tau(t,x)$ defined in Fig. 1. To explain Fig. 1 consider the transition $t \to x$ and suppose $t - \frac{d}{2} \geqslant 0$ and $t + \frac{d}{2} \leqslant 1 - d$; i.e. $\frac{d}{2} \leqslant t \leqslant 1 - \frac{3d}{2}$. By definition of the NS rule we see that if $t - \frac{d}{2} < x < t + \frac{d}{2}$, then either the request causing the transition $t \to x$ occurred at x and was served by the left head, or it occurred at $x + d$ and was served by the right head. Thus, $\tau(t,x)dx = 2dx$ for $x \in (t - \frac{d}{2}, t + \frac{d}{2})$. But if $x < t - \frac{d}{2}$ then only a request at x could have occurred; a request at $x + d$ would have been served by the *left* head moving to $x + d$, not the right head. Similarly, if $x > t + \frac{d}{2}$ only a request at $x + d$ could

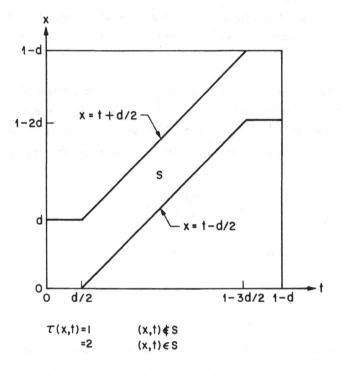

Figure 1 - Transition probability density.

have occurred. Thus, $\tau(t,x)dx = dx$ for $x \notin (t - \dfrac{d}{2}, t + \dfrac{d}{2})$. For the boundary cases $t - \dfrac{d}{2} < 0$ and $t + \dfrac{d}{2} > 1 - d$ the same argument applies *mutatis mutandis* to the intervals $(0, t + \dfrac{d}{2})$ and $(t - \dfrac{d}{2}, 1-d)$, so we obtain the result in Fig. 1. Based on this result we have

Theorem 2. Within the range $\dfrac{2}{5} \leqslant d \leqslant \dfrac{1}{2}$,

$$p(x) = \begin{cases} \dfrac{4d+1}{2} + 2x, & 0 < x < 1-2d, \\ 2, & 1-2d < x < d, \\ \dfrac{4d+1}{2} + 2(1-d-x), & d < x < 1-d, \end{cases}$$

and for $\dfrac{1}{3} \leqslant d \leqslant \dfrac{2}{5}$,

$$p(x) = \begin{cases} \dfrac{1+6d}{2} \dfrac{\sin x - \cos(1 - \frac{5d}{2} - x)}{1 - \sin(1 - \frac{5d}{2})} + 3, & 0 \leqslant x \leqslant 1 - \dfrac{5d}{2}, \\[4mm] \dfrac{1+4d}{2} + 2x, & 1 - \dfrac{5d}{2} \leqslant x \leqslant \dfrac{d}{2}, \\[4mm] \dfrac{1+6d}{2} \dfrac{\cos(x - \frac{d}{2}) - \sin(1-2d-x)}{1 - \sin(1 - \frac{5d}{2})}, & \dfrac{d}{2} \leqslant x < 1-2d, \\[4mm] 2, & 1-2d < x < d, \\[2mm] p(1-d-x), & d \leqslant x \leqslant 1-d. \end{cases}$$

Proof. Substituting for $\tau(t,x)$ in (1), $p(x)$ can now be written as the sum of two terms taken from the following expression

(63)

$$p(x) = \left\{ \begin{array}{ll} 1, & 0 < x < d, \\[2mm] \displaystyle\int_{x - \frac{d}{2}}^{1-d} p(t)dt, & d < x < 1-d, \end{array} \right\} +$$

$$\left\{ \begin{array}{ll} \displaystyle\int_{0}^{x + \frac{d}{2}} p(t)dt, & 0 < x < 1-2d, \\[4mm] 1, & 1-2d < x < 1-d. \end{array} \right\}$$

For $\dfrac{1}{3} \leqslant d \leqslant \dfrac{1}{2}$ we have $0 \leqslant 1 - 2d \leqslant d \leqslant 1 - d$ and hence (63) may be rendered as

(64)

$$p(x) = \begin{cases} 1 + \displaystyle\int_{0}^{x + \frac{d}{2}} p(t)dt, & 0 < x < 1 - 2d, \\[4mm] 2, & 1 - 2d < x < d, \\[4mm] 1 + \displaystyle\int_{x - \frac{d}{2}}^{1-d} p(t)dt, & d < x < 1 - d. \end{cases}$$

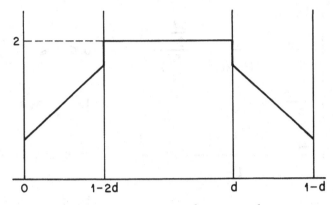

Figure 2 - The Case $\dfrac{2}{5} < d < \dfrac{1}{2}$.

Differentiating, we obtain

(65)
$$p'(x) = \begin{cases} p(x + \dfrac{d}{2}), & 0 < x < 1-2d, \\ 0, & 1-2d < x < d, \\ -p(x - \dfrac{d}{2}), & d < x < 1-d. \end{cases}$$

Since the first derivative is determined by the function evaluated at points translated a distance $d/2$, the analysis will be based on a partition of $(0,1-d)$ into sub-intervals that suitably map into each other under this translation. Such a partition depends on the relative magnitudes of $\dfrac{d}{2}$ and $1-2d$.

Case 1. $d/2 \geqslant 1-2d$, i.e. $\dfrac{2}{5} \leqslant d \leqslant \dfrac{1}{2}$. Since a translation of $\dfrac{d}{2}$ maps $(0,1-2d)$ into $(1-2d,d)$ and a translation of $-\dfrac{d}{2}$ maps $(d,1-d)$ into $(1-2d,d)$, we shall retain the partition in (65). From (64) and (65) we have $p'(x) = 2$, $0 < x < 1-2d$, and hence for some constant b we have $p(x) = b + 2x$, $0 < x < 1-2d$. It is easily seen that $p(x)$ must be symmetric on $(0,1-d)$ and therefore, as in Fig. 2, we must have

(66)
$$p(x) = \begin{cases} b + 2x, & 0 < x < 1-2d, \\ 2, & 1-2d < x < d, \\ b + 2(1-d-x), & d < x < 1-d. \end{cases}$$

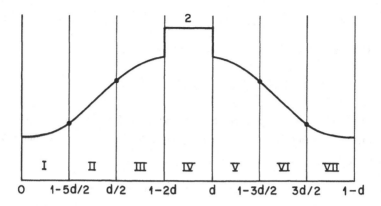

Figure 3 - The Case $\frac{1}{3} < d < \frac{2}{5}$.

To evaluate b, we may use $\int_0^{1-d} p(x)dx = 1$. This yields $b = \frac{4d+1}{2}$ and the result

of the theorem for $\frac{2}{5} \leqslant d \leqslant \frac{1}{2}$.

Case 2. $\frac{d}{2} \leqslant 1 - 2d$, i.e. $\frac{1}{3} \leqslant d \leqslant \frac{2}{5}$. We partition $(0, 1-d)$ into the 7 sub-intervals labeled $I, ..., VII$ in Fig. 3. Observe that $x \to x + \frac{d}{2}$ maps interval I onto III, III onto V, and V onto VII. Also, II is mapped onto IV and IV onto VI. As before, we use the known result from (64), viz. $p(x) = 2$ for $x \in IV$, and then apply (65) and the symmetry argument to obtain the form of $p(x)$ over the remaining intervals.

Thus, since $x \to x + \frac{d}{2}$ maps II onto IV, we have from (4) that $p'(x) = 2$, $x \in II$, and hence for some constant a_1,

(67) $$p(x) = a_1 + 2x \ , \quad x \in II \ .$$

For $x \in III$ we differentiate (65) and get $p''(x) = p'(x+\frac{d}{2}) = -p(x)$, or $p''(x) + p(x) = 0$, $x \in III$. It follows that for constants a_2 and a_3

(68) $$p(x) = a_2\cos x + a_3\sin x \ , \quad x \in III \ .$$

Since $x \rightarrow x + \dfrac{d}{2}$ maps I onto III and since $p'(x) = p(x+\dfrac{d}{2})$, $x \in I$, by (65), we can integrate (69) and obtain for some constant a_4,

(69) $$p(x) = a_2\sin(x + \frac{d}{2}) - a_3\cos(x+\frac{d}{2}) + a_4 \ , \quad x \in I \ .$$

Using the symmetry of $p(x)$ we have thus determined the form of $p(x)$ over each of the 7 intervals. Continuity of $p(x)$ in $(0,1-2d)$ and $(d,1-d)$ follows from (64), so we have determined that $p(x)$ has the form shown in Fig. 3.

To evaluate the constants $a_1 - a_4$ we may use continuity and symmetry arguments as follows. Let $a = p(\dfrac{d}{2})$. Then $p(x) = a + 2(x - \dfrac{d}{2})$ on II (see Fig. 2), and $a_1 = a - d$. By continuity the right hand side of (68) must also evaluate to a at $d/2$. From (65) we have $p'((\dfrac{d}{2})^+) = p(d^+)$. By symmetry $p(d^+) = p((1-2d)^-)$ and therefore $p'((\dfrac{d}{2})^+) = p((1-2d)^-)$. It is readily verified that the conditions $p(\dfrac{d}{2}) = a$ and $p'((\dfrac{d}{2})^+) = p((1-2d)^-)$ applied to (68) give

(70) $$p(x) = a\ \frac{\cos(x - \dfrac{d}{2}) - \sin(1-2d-x)}{1 - \sin(1 - \dfrac{5d}{2})} \ , \quad x \in III \ .$$

Integrating (70) as we did (68), (69) becomes

(71) $$p(x) = a\ \frac{\sin x - \cos(1 - \dfrac{5d}{2} - x)}{1 - \sin(1 - \dfrac{5d}{2})} + a_4, \quad x \in I \ .$$

By continuity at $x = 1 - \dfrac{5d}{2}$ we have, from (67) and (71), that $a_4 = 2a+2(1-3d)$.

It remains to find a. From (64) we have $p(0) = 1 + \displaystyle\int_0^{d/2} p(t)dt$ and $p((1-2d)^-) = 1 + \displaystyle\int_0^{1-3d/2} p(t)dt$. Using $\displaystyle\int_0^{1-d} p(t)dt = 1$ and symmetry, we obtain

$$p((1-2d)^-) = 2 - \int_{1-3d/2}^{1-d} p(t)dt = 2 - \int_0^{d/2} p(t)dt \; .$$

Thus, $p(0) + p((1-2d)^-) = 3$, so from (70) and (71) evaluated at $1 - 2d$ and 0, respectively, we get $a = \dfrac{1 + 6d}{2}$. Accumulating results we have the statement of the theorem for $\dfrac{1}{3} \leqslant d \leqslant \dfrac{2}{5}$. $\blacksquare$

The Optimization Problem — We are now in position to find that value of d minimizing $E(d)$. For $\dfrac{1}{3} < d < \dfrac{1}{2}$ we can calculate E_d in (62) by substituting for $\tau(x,t)$ according to the definition in Fig. 1 and by substituting for $p(x)$ from Theorem 1. The calculations are lengthy but essentially mechanical, and therefore omitted. The results are as follows. The integrations in (62) yield

$$\min_{\frac{2}{5} \leqslant d \leqslant \frac{1}{2}} E_d = \min_{\frac{2}{5} \leqslant d \leqslant \frac{1}{2}} \frac{2 - 8d + 33d^2 - 74d^3 + 64d^4}{6} \; ,$$

and with somewhat greater effort

$$\min_{\frac{1}{3} \leqslant d \leqslant \frac{2}{5}} E_d = \min_{\frac{1}{3} \leqslant d \leqslant \frac{2}{5}} \frac{1}{6} \left[(6+60d+72d^2-432d^3) \frac{\cos(\frac{5d}{2} - 1)}{\sin(\frac{5d}{2} - 1) + 1} \right.$$

$$\left. - 18-121d+357d^2+30d^3-216d^4 \right] \; .$$

From a numerical evaluation of these functions we find at $d = 0.44657$

(72) $$\min_{\frac{1}{3} \leqslant d \leqslant \frac{1}{2}} E_d \approx 0.16059 \; .$$

It remains to consider the case E_d, $0 < d < 1/3$. By a "bootstrapping" technique we now obtain a lower bound on $p(x)$. This gives us a corresponding lower bound on E_d, $0 < d < 1/3$, and the following result.

Theorem 3. We have

$$\min_{0 \leqslant d \leqslant \frac{1}{3}} E_d > \min_{\frac{1}{3} \leqslant d \leqslant \frac{1}{2}} E_d$$

Proof. We return to (63), which for $d < 1/3$ and hence $0 < d < 1-2d$, can be rendered as

$$
(73) \qquad p(x) = \begin{cases}
1 + \displaystyle\int_0^{x+\frac{d}{2}} p(t)\,dt\,, & 0 < x < d\,, \\[4ex]
1 + \displaystyle\int_{x-\frac{d}{2}}^{x+\frac{d}{2}} p(t)\,dt\,, & d < x < 1 - 2d\,, \\[4ex]
1 + \displaystyle\int_{x-\frac{d}{2}}^{1-d} p(t)\,dt\,, & 1 - 2d < x < 1 - d\,.
\end{cases}
$$

By (73) we have $p(t) \geqslant 1,\ 0 \leqslant t \leqslant 1 - d$, so after substitution into (73) we find that $p(x) \geqslant 1 + \dfrac{d}{2}$ for all $0 \leqslant x \leqslant 1 - d$. Substituting this bound back into (73) then gives us $p(x) \geqslant 1 + \dfrac{d}{2} + (\dfrac{d}{2})^2$, $0 \leqslant x \leqslant 1 - d$. Since this process may be continued indefinitely, it follows that

$$
p(x) \geqslant \sum_{i=0}^{\infty} (\tfrac{d}{2})^i = \frac{2}{2 - d}\,, \qquad 0 \leqslant x \leqslant 1 - d\,.
$$

Substituting this bound into each of the expressions in (74) produces

$$
(74) \qquad p(x) \geqslant \begin{cases}
\dfrac{2+2x}{2-d}\,, & 0 \leqslant x < d\,, \\[3ex]
\dfrac{2+d}{2-d}\,, & d \leqslant x < 1 - 2d\,, \\[3ex]
\dfrac{4-2d-2x}{2-d}\,, & 1 - 2d \leqslant x \leqslant 1 - d\,.
\end{cases}
$$

Omitting the tedious calculations, the corresponding bound on E_d is given by

$$
E_d \geqslant \frac{128 - 272d + 36d^2 + 480d^3 - 335d^4}{192(2-d)}\,, \qquad 0 \leqslant d \leqslant \frac{1}{3}\,.
$$

It is routine to verify that this bound is a decreasing function of d in $[0, 1/3]$. Since at $d = 1/3$ the value of the bound is $0.18741 > 0.16059$, the theorem is proved. ∎

We remark that an exact analysis for $d < 1/3$ extends the approach used for $d > 1/3$. However, for arbitrary $d < 1/3$ closed forms do not appear possible; instead, $p(x)$ is determined by a linear system in about $\lfloor 2/d \rfloor$ unknowns. The interested reader is referred to [CalCF2].

Final Remarks — In [CalCF2] a model where the heads are allowed to move off the interval [0,1] is also briefly considered. Although as noted in [PW], engineering issues may well make this proposal impractical, the methods of earlier sections are directly applicable to this case. Because of the similarity we only illustrate this variation.

First, with a fixed head separation d, the left head is now free to move in the interval $[-d,1]$. If $d \geqslant 2$ then the server initially in [0,1] will be the nearer server for all requests. Thus, $d < 2$ may be assumed. For $0 < d < 1$ an analysis paralleling that in section 2 leads to an integral equation similar to (73) with the designated intervals $(0,d)$, $(d,1-2d)$, and $(1-2d,1-d)$ replaced by $(-d,0)$, $(0,1-d)$, and $(1-d,1)$, respectively. Similarly, for $1 < d < 2$, we obtain an integral equation like (64) with the intervals $(0,1-2d)$, $(1-2d,d)$, and $(d,1-d)$ replaced by $(-d,1-d)$, $(1-d,0)$, and $(0,1)$, respectively. Solutions to these equations proceed as before. Note that the calculations for $0 < d < 1$ resemble those for the case $0 < d < 1/3$ in the original model. Thus, although the earlier approach to finding the optimal separation again applies, it may not be practical to base the computations on closed forms.

It is interesting to observe that expected head motion under the NS rule in the system [CalCF1] where the two heads move independently but one at a time, is greater than in the system analyzed here. It is 0.1625 in the former system and 0.1606 here. We note also the interesting property that in terms of expected head motion under the NS rule both systems perform *more* than twice better than a single head system, where expected head motion is 1/3.

It is of interest to compare the performance of disk systems with two heads a fixed distance d apart under the nearer-server policy with their performance under an optimal head-selection policy. As an experiment a disk system with 201 storage addresses spaced equally along the unit interval was considered in [CalCF2]. The head separation d was taken to be $i/200$ and i was allowed to range between 70

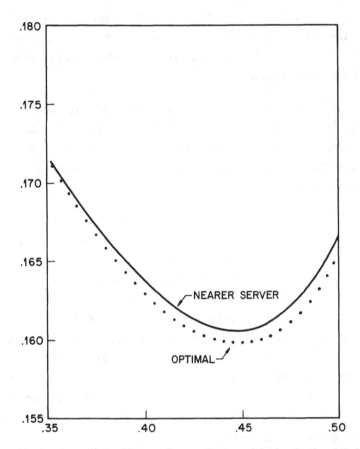

Figure 4 - Comparison of the Nearer Server Policy with the Optimal Policy.

and 100. For a fixed head separation $d = i/200$ calculations were made of the expected cost of servicing 500 requests under an optimal head-selection policy (an average was taken over the 201-i possible starting positions for the two heads). Figure 4 compares the expected cost per request under the optimal policy with the expected head motion under the nearer-server rule. The difference in performance is very small but it does appear that the nearer-server rule is not optimal.

Note that it is possible to calculate recursively the expected cost of servicing N requests. For suppose that for all j we know the expected cost $C_N(j)$ of servicing N requests given that the left head starts in position $j/200$. Then consider servicing $N+1$ requests given that the left head is initially at position $k/200$. Since the two heads are a distance $i/200$ apart we have

$$201\,C_{N+1}(k) = \sum_{\ell=0}^{i-1} [\frac{|k-\ell|}{100}+C_N(\ell)] + \sum_{\ell=200-i+1}^{200} [\frac{|k+i-\ell|}{100}+C_N(\ell-i)]$$
$$+ \sum_{\ell=i}^{200-i} \min(\frac{|k-\ell|}{100}+C_N(\ell), \frac{|k+i-\ell|}{100}+C_N(\ell-i)) .$$

This was the recurrence used in [CalCF2] to calculate the expected unit cost of servicing 500 requests under an optimal policy.

References

[AbuKL] Abu-Sufah, W., D. J. Kuck and D. H. Lawrie, "On the Performance Enhancement of Paging Systems Through Program Analysis and Transformations," *IEEE Trans. Comput.*, **30** (1981), 341-356.

[AhoDU] Aho, A. V., P. J. Denning and J. D. Ullman, "Principles of Optimal Page Replacement," *J. Assoc. Comput. Mach.*, **18** (1971), 80-93.

[AniMS] Anick, D., D. Mitra and M. M. Sondhi, "Stochastic Theory of a Data Handling System with Multiple Sources," *Bell Sys. Tech. J.*, **61** (1982), 1871-1894.

[ArrKS] Arrow, K. J., S. Karlin, and H. Scarf, "*Studies in the Mathematical Theory of Inventory and Production*," Stanford University Press, Stanford, CA, 1958.

[AveBK1] Aven, O. I., L. B. Boguslavsky, and Y. A. Kogan, "Some Aspects of the Non-Parametric Theory of Replacement Algorithms," *Proc. Int. Symp. on Comp. Perf. Evaluation*, E. Gelenbe and D. Potier (eds.), North Holland, Amsterdam, 1975, 226-231.

[AveBK2] Aven, O. I., L. B. Boguslavsky, and Y. A. Kogan, "Some Results on Distribution-Free Analysis of Paging Algorithms," *IEEE Trans. Comput.*, **25** (1976), 737-745.

[AveGK] Aven, O. I., N. N. Gurin and Y. A. Kogan, *Performance Evaluation and Optimization of Computer Systems*, Nauka, Moscow, 1982 [in Russian].

236

[AveK1] Aven, O. I. and Y. A. Kogan, "On a Stochastic Control Problem Concerning Virtual Memory: The Comparison of Heuristic and Optimal Paging Algorithms," *Proc. IFAC Symp. on Stoch. Cont.*, Budapest, 1974, 481-490.

[AveK2] Aven, O. I. and Y. A. Kogan, "Stochastic Control of Paging in a Two-Level Computer Memory," *Automatica*, **11** (1975), 309-313.

[AveK3] Aven, O. I. and Y. A. Kogan, *"Computer Resource Allocation: Algorithms and Models,"* Energiya, Moscow, 1978 [in Russian].

[AveS] Aven, O. I. and V. B. Sokolov, "On Some Methods of Memory Management," *Technical Cybernetics*, **2** (1971), 89-94 [in Russian].

[BabF] Babaoglu, O. and D. Ferrari, "Two Level Replacement Decisions in Paging Stores," *IEEE Trans. Comput.*, **32** (1983), 1151-1159.

[BarS] Barrese, A. L. and S. D. Shapiro, "Structuring Programs for Efficient Operation in Virtual Memory Systems," *IEEE Trans. Soft. Eng.*, **5** (1979), 643-652.

[Bela] Belady, L. A., "A Study of Replacement Algorithms for Virtual Storage Computers," *IBM Sys. J.*, **5** (1966), 78-101.

[BelaPS] Belady, L. A., R. P. Parmallee and C. A. Scalzi, "The IBM History of Memory Management Technology," *IBM J. Res. and Devel.*, **25** (1981), 491-503.

[Bell] Bellman, R., *Introduction to Matrix Analysis*, 2nd McGraw-Hill, New York, 1970, p. 93.

[Ben1] Beneš V. E., "Storage Problems Involving Finite Capacity, Loading Protocols and Exact Content," AT&T Bell Laboratories, Murray Hill, N.J. 07974, 1984 (to appear).

[Ben2] Beneš, V. E., "Optimal Memory Allocation: Problems, Principles and an Example," AT&T Bell Laboratories, Murray Hill, N. J. 07974, 1983 (to appear).

[Bil] Billingsley, P., *Convergence of Probability Measures*, John Wiley & Sons, New York, 1968.

[BogF] Bogott, R. P. and M. A. Franklin, "Evaluation of Markov Program Models in Virtual Memory Systems," *Soft. Prac. and Exp.*, **5** (1975), 337-346.

[BogK] Boguslavsky, L. B. and Y. A. Kogan, "An Analysis of Page Replacement Algorithms for Two-Level Memory," *Auto. Remote Cont.*, **35** (1974), Pt. 2, 1818-1824.

[Bor] Borovkov, A. A., "On Limit Laws for Service Processes in Multi-Channel Systems," *Siberian Math. J.*, **8** (1967), 746-763 [English translation].

[Bro] Bronshtein, I. I., "Optimal Strategies for Swapping Equal Length Programs in and out of Computer Operational Memory," *Auto. Remote Cont.*, **33** (1972), Pt. 2, 452-458.

[BroK] Bronshtein, I. I. and Y. A. Kogan, "Comparison of Sub-Optimal and Optimal Replacement Algorithms in a Two-Level Memory: Numerical Results," *Auto. Remote Cont.*, **35** (1974), Pt. 2, 2014-2016.

[BudDMS] Budzinski, R. I., E. S. Davidson, W. Mayeda and H. S. Stone, "DMIN: An Algorithm for Computing the Optimal Dynamic Allocation in a Virtual Memory Computer," *IEEE Trans. Soft. Eng.*, **7** (1981), 113-120.

[CalCF1] Calderbank, A. R., E. G. Coffman, Jr. and L. Flatto, "Sequencing Problems in Two Server Systems," *Math. Oper. Res.* (in press).

[CalCF2] Calderbank, A. R., E. G. Coffman, Jr. and L. Flatto, "Optimum Head Separation in a Disk System with Two Read/Write Heads," *J. Assoc. Comput. Mach.*, **31** (1984), 826-838.

[ChaW] Chandra, A. K. and C. K. Wong, "Worst-Case Analysis of a Placement Algorithm Related to Storage Allocation," *SIAM J. Comput.*, **4** (1975), 249-263.

[ChuL] Chu, W. W. and L. C. Liang, "Buffer Behavior for Mixed Input Traffic and a Single Constant Output Rate," *IEEE Trans. Commun.*, **COM-20** (1972), 230-235.

[ChuO] Chu, W. W. and H. Opderbeck, "Analysis of the PFF Algorithm Using a Semi-Markov Model," *Comm. Assoc. Comput. Mach.*, **19** (1976), 298-304.

[ChusH] Chusho, T. and T. Hayashi, "Performance Analysis of Paging Algorithms for Compilation of a Highly Modularized Program," *IEEE Trans. Soft. Eng.*, **7** (1981), 248-254.

[Cof1] Coffman, E. G., Jr., "Analysis of a Drum Input/Output Queue under Scheduled Operation in a Paged Computer System," *J. Assoc. Comput. Mach.*, **16** (1969), 73-90 [Corrigendum: **16** (1969), 646].

[CofD] Coffman, E. G., Jr. and P. J. Denning, *Operating Systems Theory*, Prentice-Hall, Englewood Cliffs, N.J., 1973.

[CofGJ] Coffman, E. G., Jr., M. R. Garey and D. S. Johnson, "Approximation Algorithms for Bin-Packing-An Updated Survey," *Algorithm Design for Computer System Design*, G. Ausiello, M. Lucertini, and P. Serafini (eds.), Springer-Verlag, New York, 1984, 49-106.

[CofH1] Coffman, E. G., Jr. and M. Hofri, "A Class of FIFO Queues Arising in Computer Systems," *Oper. Res.*, **26** (1978), 864-880.

[CofH2] Coffman, E. G., Jr. and M. Hofri, "On the Expected Performance of Scanning Disks," *SIAM J. Comput.*, **11** (1982), 60-70.

[CofKS1] Coffman, E. G., Jr., T. T. Kadota and L. A. Shepp, "A Stochastic Model of Fragmentation in Dynamic Storage Allocation," *SIAM J. Comput.*, **14** (1985), 416-425.

[CofKS2] Coffman, E. G., Jr., T. T. Kadota and L. A. Shepp, "On the Asymptotic Optimality of First-Fit Storage Allocation," *IEEE Trans. Soft. Eng.*, **SE-11** (1985), 235-239.

[CofL] Coffman, E. G., Jr. and F. T. Leighton, "A Provably Efficient Algorithm for Dynamic Storage Allocation," *Proc. ACM Symp. on Th. of Comput.*, May 1986, 77-90.

[CofR] Coffman, E. G., Jr. and M. I. Reiman, "Diffusion Approximations for Storage Processes in Computer Systems," *Proc. Int. Workshop Appl. Math. and Performance/Reliability Models of Comput./Commun. Sys.*, G. Iazeolla and S. Tucci (eds.), North Holland, 1983, 33-54.

[Coh1] Cohen, J. W., *The Single Server Queue*, North Holland, Amsterdam, 1969.

[Coh2] Cohen, J. W., "Superimposed Renewal Processes and Storage with Gradual Input," *Stoch. Proc. and Their Appl.*, **2** (1974), 31-58.

[Coo] Cooper, R. B., *Queueing Theory*, Macmillan, New York, 1972.

[Cou] Courtois, P. J., *Decomposability*, Academic Press, New York, 1977.

[CouL] Courtois, P. J. and G. Louchard, "Approximation of Eigencharacteristics in Nearly Completely Decomposable Stochastic Systems," *Stoch. Proc. and Their Appl.*, **4** (1976), 283-296.

[CouS] Courtois, P. J. and P. Semal, "Error Bounds for the Analysis by Decomposition of Non-negative Matrices," *Proc. Int. Workshop Appl. Math. and Performance/Reliability Models of Comput./Commun. Sys.*, G. Iazeolla and S. Tucci (eds.), North-Holland, 1983, 253-268.

[CouV] Courtois, P. J. and H. Vantilborgh, "A Decomposable Model of Program Paging Behavior," *Acta Informatica*, **6** (1976), 251-275.

[CoxM] Cox, D. R. and H. D. Miller, *The Theory of Stochastic Processes*, John Wiley & Sons, New York, 1965.

[Den1] Denning, P. J., "The Working Set Model of Program Behavior," *Comm. Assoc. Comput. Mach.*, **11** (1968), 323-333.

[Den2] Denning, P. J., "Virtual Memory," *Comput. Surv.*, **2** (1970), 153-189.

[Den3] Denning, P. J., "Working Sets: Past and Present," *IEEE Trans. Soft. Eng.*, **6** (1980), 64-84.

[DenK] Denning, P. J. and K. C. Kahn, "A Study of Program Locality and Lifetime Functions," *Oper. Sys. Rev.*, **9** (1975), 207-216.

[DenSl] Denning, P. J. and D. R. Slutz, "Generalized Working Sets for Segment Reference Strings," *Comm. Assoc. Comput. Mach.*, **21** (1978), 750-759.

[DenSc] Denning, P. J. and S. C. Schwartz, "Properties of the Working Set Model," *Comm. Assoc. Comput. Mach.*, **15** (1972), 191-198.

[Der] Derman, C., *Finite State Markovian Decision Processes*, Academic Press, New York, 1970.

[Die] Dies, J. E., *Chaines de Markov sur les Permutations,* Lecture Notes in Mathematics 1010, A. Dold and B. Eckmann (eds.), Springer-Verlag, Berlin, 1983.

[Eas1] Easton, M. C., "A Model for Interactive Data Base Reference Strings," *IBM J. Res. and Devel.*, **19** (1975), 550-556.

[Eas2] Easton, M. C., "A Model for Data Base Reference Strings Based on Behavior of Reference Clusters," *IBM J. Res. and Devel.*, **22** (1978), 197-202.

[EasW] Easton, M. C. and C. K. Wong, "The Effect of a Capacity Constraint on the Minimal Cost of a Partition," *J. Assoc. Comput. Mach.*, **22** (1975), 441-449.

[Eis] Eisenberg, M., "Queues with Periodic Service and Changeover Time," *Oper. Res.*, **20** (1972), 440-451.

[Fag] Fagin, R., "Asymptotic Miss Ratios over Independent References," *J. Comp. Sys. Sci.*, **14** (1977), 222-250.

[FagE] Fagin, R. and M. S. Easton, "The Independence of Miss Ratio on Page Size," *J. Assoc. Comput. Mach.*, **23** (1976), 128-146.

[FagP] Fagin, R. and T. G. Price, "Efficient Calculation of Expected Miss Ratios in the Independent Reference Model," *SIAM J. Comput.*, **7** (1978), 288-296.

[Fell] Feller, W., *An Introduction to Probability Theory and Its Applications*, John Wiley & Sons, New York, Vol. I, 3rd Ed., 1968, Vol. II, 1966.

[Fel2] Feller, W., "Diffusion Processes in One Dimension," *Trans. Amer. Math. Soc.*, **77** (1954), 1-31.

[FerLW] Fernandez, E. G., T. Lang and C. Wood, "Effect of Replacement Algorithms on a Paged Buffer Data Base System," *IBM J. Res. and Devel.*, **22** (1978), 185-196.

[FoxL] Fox, B. L. and D. M. Landi, "An Algorithm for Identifying the Ergodic Subchains and Transient States of a Stochastic Matrix," *Comm. Assoc. Comput. Mach.*, **11** (1968), 619-621.

[FranaW] Franascek, P. A. and T. J. Wagner, "Some Distribution-Free Aspects of Paging Algorithm Performance," *J. Assoc. Comput. Mach.*, **21** (1974), 31-39.

[Frank] Frank, H., "Analysis and Optimization of Disk Storage Devices for Time-Sharing Systems," *J. Assoc. Comput. Mach.*, **16** (1969), 602-620.

[FrankG] Franklin, M. A. and R. K. Gupta, "Computation of PF Probabilities from Program Transition Diagrams," *Comm. Assoc. Comput. Mach.*, **17** (1974), 186-191.

[FreW] Freidlin, M. I. and A. D. Wentzell, *Random Perturbations of Dynamical Systems*, Springer-Verlag, Berlin, 1984.

[FulB] Fuller, S. H. and F. Baskett, "An Analysis of Drum Storage Units," *J. Assoc. Comput. Mach.*, **22** (1975), 83-105.

[Gan] Gani, J., "Problems in the Probability Theory of Storage Systems," *J. Roy. Stat. Soc., B*, **19** (1957), 181-206.

[GanP] Gani, J. and N. U. Prabhu, "Remarks on the Dam with Poisson Type Inputs," *Australian J. Appl. Sci.*, **10** (1959), 113-122.

[GavL] Gaver, D. P., Jr. and P. A. W. Lewis, "Probability Models for Buffer Storage Allocation Problems," *J. Assoc. Comput. Mach.*, **18** (1971), 186-198.

[GavM] Gaver, D. P., Jr. and R. G. Miller, Jr., "Limiting Distributions for Some Storage Problems," in *Studies in Appl. Prob. and Manag. Sci.*, K. J. Arrow, S. Karlin and H. Scarf (eds.), Stanford University Press, 1962, 110-126.

[Gel1] Gelenbe, E., "A Unified Approach to the Evaluation of a Class of Replacement Algorithms," *IEEE Trans. Comput.*, **22** (1973), 611-618.

[Gel2] Gelenbe, E., "On Approximate Computer System Models," *J. Assoc. Comput. Mach.*, **22** (1975), 261-269.

[GelTB] Gelenbe, E., P. Tiberio and J. C. A. Bockhoerst, "Page Size in Demand Paging Systems," *Acta Informatica*, **3** (1973), 1-23.

[GelM] Gelenbe, E. and I. Mitrani, *Analysis and Synthesis of Computer Systems*, Academic Press, New York, 1980.

[Glo] Glowacki, C., "A Closed Form Expression of the Page Fault Rate for the LRU Algorithm in a Markovian Reference Model of Program Behavior," *Proc. Int. Comput. Symp.*, E. Morlet and D. Ribbens (eds.), North-Holland, Amsterdam, 1977, 315-318.

[GotM] Gotlieb, C. C. and G. H. McEwan, "Performance of a Movable-Head Disk Storage Device," *J. Assoc. Comput. Mach.*, **20** (1973), 604-623.

[GupF] Gupta, R. K. and M. A. Franklin, "Working Set and Page Fault Frequency Paging Algorithms: A Performance Comparison," *IEEE Trans. Comput.*, **27** (1978), 706-712.

[GurK] Gurin, N. N. and Y. A. Kogan, "On the Organization of Two-Level Access to Files of an Operating System," *Auto. Remote Cont.*, **36** (1975), No. 12, Pt. 2.

[Hans] Hansen, P. B., *The Architecture of Concurrent Programs*, Prentice-Hall, Englewood Cliffs, N.J., 1977.

[Hand] *Handbook of Mathematical Functions*, M. Abramowitz and I. A. Stegun (Eds.), Nat. Bur. of Stand., Appl. Math. Ser. 55, 1964.

[HasF] Hashida O. and M. Fujika, "Queueing Models for Buffer Memory in Store-and-Forward Systems," *Proc. Int. Teletraffic Cong.*, Stockholm, June 1973, 323/1, 323/7.

[Hat] Hatfield, D. J., "Experiments on Page Size, Program Access Patterns and Virtual Memory Performance," *IBM J. Res. and Devel.*, **16** (1972), 58-66.

[HatG] Hatfield, D. J. and J. Gerald, "Program Restructuring for Virtual Memory," *IBM Sys. J.*, **10** (1971), 168-192.

[HavH] Haviv, M. and L. Van der Heyden, "Perturbation Bounds for a Finite Markov Chain," *Adv. Appl. Prob.*, **16** (1984), 804-817.

[Hof1] Hofri, M., "Disk Scheduling: FCFS vs. SSTF Revisited," *Comm. Assoc. Comput. Mach.*, **23** (1980), 11.

[Hof2] Hofri, M., "Should the Two-Headed Disk Be Greedy?" *Inf. Proc. Letters*, **16** (1983), 83-86.

[HofT] Hofri, M. and P. Tzelnic, "The Working Set Size Distribution for the Markov Chain Model of Program Behavior," *SIAM J. Comput.*, **11** (1982), 453-466.

[How] Howard, R. A., *Dynamic Programming and Markov Processes*, M.I.T. Technology Press, Cambridge, Mass., 1960.

[IngK] Ingargiola, G. and J. F. Korsh, "Finding Optimal Demand Paging Algorithms," *J. Assoc. Comput. Mach.*, **21** (1974), 40-53.

[Igl1] Iglehart, D. L., "Limit Diffusion Approximations for the Many-Server Queue and the Repairman Problem," *J. Appl. Prob.*, **2** (1965), 429-441.

[Igl2] Iglehart, D. L., "Weak Convergence of Compound Stochastic Processes," *Stoch. Proc. Appl.*, **1** (1973), 11-31.

[IglW] Iglehart, D. L. and W. Whitt, "Multiple Channel Queues in Heavy Traffic, I and II," *Adv. Appl. Prob.*, **2** (1970), 150-177 and 355-364.

[Jac1] Jackson, J. R., "Networks of Waiting Lines," *Oper. Res.*, **15** (1957), 254-265.

[Jac2] Jackson, J. R., "Jobshop-Like Queueing Systems," *Manag. Sci.*, **10** (1963), 131-142.

[KanR] Kan, Y. C. and S. M. Ross, "Optimal List Order under Partial Memory Constraints," *J. Appl. Prob.*, **17** (1980), 1004-1015.

[KeiR] Keilson, J. and H. F. Ross, "First Passage Time Distributions for Gaussian Markov (Ornstein-Uhlenbeck) Statistical Processes," *Selected Tables in Math. Stat.*, **3** (1975), 233-327.

[Kel] Kelly, F. P., *Reversibility and Stochastic Networks*, John Wiley & Sons, New York, 1979.

[KemS] Kemeny, J. G. and J. L. Snell, *Finite Markov Chains*, Van Nostrand, Princeton, N.J., 1960.

[Ken] Kendall, D. G., "Some Problems in the Theory of Dams," *J. Roy. Stat. Soc., B*, **19** (1957), 207-212.

[Kin] King, W. F., III, "Analysis of Paging Algorithms," *Proc. IFIP Cong.*, Lyublyana, Yugoslavia, Aug. 1971, 485-490.

[Kle] Kleinrock, L. *Queueing Systems, Vol. 1 Theory*, John Wiley & Sons, New York, 1975.

[Knu] Knuth, D. E., *Sorting and Searching, The Art of Computer Programming*, Vol. 3, Addison-Wesley, Reading, Mass., 1973.

[Kog1] Kogan, Y. A., "Markov Models of Page Replacement in a Two-Level (Virtual) Computer Memory," *Auto. Remote Cont.*, **34** (1973), Pt. 2, 641-648.

[Kog2] Kogan, Y. A., "Analytic Investigation of Replacement Algorithms for Markov and Semi-Markov Program Behavior Models," *Auto. Remote Cont.*, **38** (1977), Pt. 2, 109-111.

[Kog3] Kogan, Y. A., "A General Approach to Page Replacement Control in a Two-Level Memory," *Proc. All Union Conf. on Control Sci.*, ITK, Minsk, 1977, 444-447 [in Russian].

[Kog4] Kogan, Y. A., "A Class of Hierarchical Paging Algorithms," *Selected Papers on Operating Systems*, M. Arato and E. Knuth (eds.), KSH SZAMKI, Budapest, 1978, 239-244.

[Kog5] Kogan, Y. A., "Problems of I/O Sub-System Optimization in High Speed Computers," *Proc. 2nd All Union Conf. on High Speed Computers*, Moscow, 1984, 152-153 (in Russian).

[Kog6] Kogan, Y. A., "On Modeling and Performance Evaluation of Programs with Overlapping Localities," *Auto. Remote Cont.*, **41** (1980), No. 12, Pt. 2.

[KogL] Kogan, Y. A. and R. S. Liptser, "Gaussian Diffusion Approximations for Closed Queueing Networks," *Fundamentals of Teletraffic Theory*, Proc. 3-rd Int. Sem. Teletraffic Theory, Moscow, 1984, 244-252.

[KogT] Kogan, Y. A. and G. G. Torosyan, "On Nearly Decomposable Markov Chains and Their Applications," *Problems of Information Transmission*, **4** (1980).

[Kos1] Kosten, L., "Uber Sperrungswahrsheinlichkeiten bei Staffelschaltungen," *Electra Nachrichten*, **14** (1937), 5-12.

[Kos2] Kosten, L., "Stochastic Theory of a Multi-Entry Buffer (1)," Delft Progress Report, **1** (1974), 10-18.

[Kos3] Kosten, L., *Stochastic Theory of Service Systems*, Pergamon Press, 1973.

[Let] Letac, G., *Chaines de Markov sur les Permutations*, SMS Presses de l'Université de Montréal, 1978.

[MadB] Madison, A. W. and A. P. Batson, "Characteristics of Program Locality," *Comm. Assoc. Comput. Mach.*, **19** (1976), 285-294.

[MatGST] Mattson, R. L., J. Gecsei, D. R. Slutz, and I. L. Traiger, "Evaluation Techniques for Storage Hierarchies," *IBM Sys. J.*, **9** (1970), 78-117.

[MicS] Michnovsky, S. D. and N. Z. Shor, "Evaluation of a Minimal Number of Transfers under Dynamic Allocation of Paging Memory," *Kibernatika*, **5** (1965), 18-20 [in Russian].

[Mit] Mitra, D., "Some Aspects of Hierarchical Memory Systems," *J. Assoc. Comput. Mach.*, **21** (1974), 54-65.

[New] Newell, G. F., *The M/M/∞ Service System with Ranked Servers in Heavy Traffic*, Springer-Verlag, New York, Lecture Notes in Economics and Mathematical Systems, 1984.

[One] Oney, W. C., "Queueing Analysis of the SCAN Policy for Movable Head Disks," *J. Assoc. Comput. Mach.*, **22** (1975), 397-412.

[OpdC] Opderbeck, H. and W. W. Chu, "The Renewal Model for Program Behavior," *SIAM J. Comput.*, **4** (1975), 356-374.

[PagW] Page, I. P. and R. T. Wood, "Emperical Analysis of a Moving Head Disc Model with Two Heads Separated by a Fixed Number of Tracks," *Comp. J.*, **24** (1981), 339-341.

[ParF] Paris, J. F. and D. Ferrari, "An Analytical Study of Strategy Oriented Restructuring Algorithms," *Performance Eval.*, **4** (1984), 117-132.

[PriF] Prieve, B. G. and R. S. Fabry, "VMIN-An Optimal Variable Space Page Replacement Algorithm," *Comm. Assoc. Comp. Mach.*, **19** (1976), 295-297.

[Pro] Prohorov, Y. V., "Convergence of Random Processes and Limit Theorems in Probability Theory," *Theor. Prob. Appl.*, **1** (1956), 157-214.

[Rao] Rao, G. S., "Performance Analysis of Cache Memories," *J. Assoc. Comput. Mach.*, **25** (1978), 378-395.

[Rei1] Reiman, M. I., "Some Diffusion Approximations with State-Space Collapse," *Modeling and Performance Evaluation Methodology*, Lecture Notes in Control and Information Sciences, F. Baccelli and G. Fayolle (eds.), Springer-Verlag, Berlin, 1983.

[Rei2] Reiman, M. I., "Multi-Class Feedback Queues in Heavy Traffic," AT&T Bell Laboratories, Murray Hill, N.J. (to appear).

[Rud] Rudin, H., "Buffered Packet Switching: A Queue with Clustered Arrivals," *Int. Switching Symp. Rec.*, M.I.T., Cambridge, Mass., June 1972.

[Sen] Sengupta, B., "The Spatial Requirement of an M/G/1 Queue, or How to Design for Buffer Space," *Modeling and Performance Evaluation Methodology*, Lecture Notes in Control and Information Sciences, F. Baccelli and G. Fayolle (eds.), Springer-Verlag, Berlin, 1983, 547-564.

[SheS] Shedler, G. S. and D. R. Slutz, "Derivation of Miss Ratios for Merged Access Streams," *IBM J. Res. and Devel.*, **20** (1976), 505-517.

[SheT] Shedler, G. S. and C. Tung, "Locality in Page Reference Strings," SIAM J. Comput., **1** (1972), 218-241.

[Ski] Skinner, C. E., "A Priority Queueing System with Server Walking Time," *Oper. Res.*, **15** (1967), 278-285.

[Smi1] Smith, A. J., "A Modified Working Set Paging Algorithm," *IEEE Trans. Comput.*, **25** (1976), 907-917.

[Smi2] Smith, A. J., "Analysis of the Optimal Lookahead Demand Paging Algorithms," *SIAM J. Comput.*, **5** (1976), 743-757.

[Smi3] Smith, A. J., "Bibliography on Paging and Related Topics," *Oper. Sys. Rev.*, **12** (1978), 39-56.

[Smi4] Smith, A. J., "Cache Memories," *Comput. Surv.*, **14** (1982), 473-530.

[Smi5] Smith, A. J., "Optimization of I/O Systems by Cache Disks and File Migration: A Summary," *Performance Eval.*, **1** (1981), 249-262.

[Spi] Spirn, J. R., *Program Behavior: Models and Measurements*, Elsevier, New York, 1977.

[SpiD] Spirn, J. R. and P. J. Denning, "Experiments with Program Locality," *Proc. AFIPS FJCC*, **41**, AFIPS Press, Montvale, N.J., 1972, 611-621.

[Ste1] Stewart, G. W., "Computable Error Bounds for Aggregated Markov Chains," *J. Assoc. Comput. Mach.*, **30** (1983), 271-285.

[Ste2] Stewart, G. W., "On the Structure of Nearly Uncoupled Markov Chains," *Proc. Int. Workshop Appl. Math. and Performance/Reliability Models of Comput./Commun. Sys.*, G. Iazeolla and S. Tucci (eds.), North-Holland, 1983, 235-251.

[SweH] Sweet, A. L. and J. C. Hardin, "Solutions for Some Diffusion Processes with Two Barriers," *J. Appl. Prob.*, **7** (1970), 423-431.

[Tak] Takács, L. "Investigation of Waiting Time Problems by Reduction to Markov Processes," *Acta Math. Acad. Sci. Hung.*, **6** (1955), 101-129.

[Tze] Tzelnic, P., "The Length of Path for Finite Markov Chains and Its Application to Modeling Program Behavior and Interleaved Memory Systems," *Proc. ORSA-TIMS Symp. Appl. Prob. and Comput. Sci.-The Interface*, Birkhäuser, Cambridge, Mass., 1981.

[TzeG] Tzelnic, P. and I. Gertner, "An Approach to Program Behavior Modeling and Optimal Memory Control," *J. Assoc. Comput. Mach.*, **29** (1982), 527-554.

[Van1] Vantilborgh, H., "On the Working Set Size Distribution and Its Normal Approximation," *BIT*, **14** (1974), 240-251.

[Van2] Vantilborgh, H., "Aggregation with an error of $O(\epsilon^2)$," *J. Assoc. Comput. Mach.*, **32** (1985), 162-190.

[VolMHKR] Voldman, J., B. Mandelbrot, L. W. Hoevel, J. Knight and P. Rosenfeld, "Fractal Nature of Software-Cache Interaction," *IBM J. Res. and Develop.*, **27** (1983), 164-170.

[Wat] Waters, S. J., "Estimating Magnetic Disk Seeks," *Comp. J.*, **18** (1975), 12-17.

[Whi] Whitt, W., "On the Heavy Traffic Limit Theorem for GI/G/∞ Queues," *Adv. Appl. Prob.*, **14** (1982), 171-190.

[Wim] Wimbrow, J. H., "A Large Scale Interactive Administration System," *IBM Sys. J.*, **10** (1971), 260-282.

[Won] Wong, C. K., *Algorithmic Studies in Mass Storage Systems*, Computer Science Press, Rockville, Md., 1983.

[Wor] Worvic, M., "Operating Systems J Level," *Executive Program and Operating Systems*, G. Cuttle and P. B. Robinson (eds.) MacDonald, London and American Elsevier, New York, 1970.

[Yag] Yaglom, A. M., *Stationary Random Functions*, Prentice-Hall, Englewood, N.J., 1962.

[YueW] Yue, P. C. and C. K. Wong, "On the Optimality of a Probability Ranking Scheme in Storage Applications," *J. Assoc. Comp. Mach.*, **20** (1973), 624-633.

[Zip] Zipf, G. K., *Human Behavior and the Principle of Least Effort*, Addison-Wesley, Reading, Mass., 1949.

INDEX